T0312807

Power Electronics

A First Course

Power Electronics

A First Course

Simulations and Laboratory Implementations

Second Edition

NED MOHAN AND SIDDHARTH RAJU
University of Minnesota, Minneapolis, MN, USA

WILEY

Library of Congress Cataloging-in-Publication Data
Names: Mohan, Ned, author. | Raju, Siddharth, author.
Title: Power electronics A first course : simulations and laboratory implementations /
    Ned Mohan and Siddharth Raju, University of Minnesota, Minneapolis, MN.
Description: Second edition. | Hoboken, New Jersey : John Wiley & Sons, [2023] |
    Includes bibliographical references and index.
Identifiers: LCCN 2022038114 (print) | LCCN 2022038115 (ebook) |
    ISBN 9781119818564 (hardback) | ISBN 9781119818571 (pdf) | ISBN 9781119818588 (epub)
Subjects: LCSH: Electric power systems. | BISAC: TECHNOLOGY & ENGINEERING /
    Power Resources / General.
Classification: LCC TK1001 .M5985 2023 (print) | LCC TK1001 (ebook) |
    DDC 621.31--dc23/eng/20221017
LC record available at https://lccn.loc.gov/2022038114
LC ebook record available at https://lccn.loc.gov/2022038115

Cover image: © Mr. Kosal/Shutterstock
Cover design: Wiley

Set in 10/12pt TimesNewRomanMTStd by Integra Software Services Pvt. Ltd, Pondicherry, India

SKY10066512_020724

To our families

CONTENTS

LIST OF SIMULATION AND HARDWARE IMPLEMENTATION EXAMPLE AND FIGURES

Experiment	Simulation	Hardware
Lab kit	N/A	Figure 2.15
Buck converter - CCM	Example 3.4 Figures 3.7 and 3.8	Example 3.4 Figures 3.9 and 3.10
Boost converter - CCM	Example 3.6 Figures 3.15 and 3.16	Example 3.6 Figures 3.17 and 3.18
Buck-Boost converter - CCM	Example 3.8 Figures 3.23 and 3.24	Example 3.8 Figures 3.25 and 3.26
Synchronous-rectified Buck converter	Example 3.9 Figures 3.30 and 3.31	Example 3.9 Figures 3.32 through 3.34
Buck converter - DCM	Example 3.10 Figures 3.43	Example 3.10 Figures 3.44
Boost converter - DCM	Example 3.12 Figures 3.47	Example 3.12 Figures 3.48
Buck-Boost converter - DCM	Example 3.13 Figures 3.50	Example 3.13 Figures 3.51
Buck converter frequency response	Example 4.2 Figures 4.9 and 4.10	N/A
Buck converter - Voltage mode control	Example 4.4 Figures 4.13 through 4.18	Example 4.4 Figures 4.19 and 4.20
Buck-Boost converter – Peak-current-mode control	Example 4.6 Figures 4.23 through 4.27	Example 4.6 Figures 4.28 and 4.29
Single-phase diode-bridge rectifier	Example 5.2 Figures 5.13 and 5.14	N/A

(Continued)

(Continued)

Experiment	Simulation	Hardware
Three-phase diode-bridge rectifier	Example 5.3 Figures 5.18 and 5.19	N/A
Control of Power-Factor-Correction circuit	Figures 6.8 and 6.9	N/A
Flyback converter – CCM without snubber	Example 8.2 Figures 8.4 and 8.5	Example 8.2 Figures 8.6 and 8.7
Flyback converter – CCM with snubber	Example 8.5 Figures 8.11 and 8.12	Example 8.5 Figure 8.13
Flyback converter – DCM without snubber	Example 8.6 Figure 8.14	Example 8.6 Figures 8.15
Forward converter	Example 8.8 Figures 8.20 and 8.21	Example 8.8 Figures 8.22 and 8.23
PWM Full-Bridge converter	Example 8.10 Figures 8.31 and 8.32	Example 8.10 Figures 8.33 and 8.34
Single-phase Inverter	Example 12.5 Figures 12.21 and 12.22	Example 12.5 Figures 12.23 and 12.24
Three-phase Inverter – Sine PWM	Example 12.7 Figures 12.29 and 12.30	Example 12.7 Figures 12.31 and 12.32
Three-phase Inverter – SVPWM	Example 12.9 Figures 12.38 and 12.39	Example 12.9 Figures 12.40 and 12.41

PREFACE

Role of Power Electronics in Providing Sustainable Electric Energy

As discussed in the introductory chapter of this textbook, power electronics is an enabling technology for powering information technology and making factory automation feasible. In addition, power electronics has a crucial role to play in providing sustainable electric energy. Most scientists now believe that carbon-based fuels for energy production contribute to climate change, which is a serious threat facing human civilization. In the United States, the Department of Energy reports that approximately 40% of all the energy consumed is first converted into electricity. Potentially, use of electric and plug-in hybrid cars, high-speed rails, and so on, may increase this to even 60%. Therefore, it is essential that we generate electricity from renewable sources such as wind and solar, which at present represent only slightly over 4%, build the next-generation smarter and robust grid to utilize renewable resources often in remote locations, and use electricity in more energy-efficient ways. Undoubtedly, using electricity efficiently and generating it from renewable sources are the twin pillars of sustainability, and as described in this textbook, power electronic systems are a key to them both!

This textbook focuses on *power electronic systems* as one of the topics in an integrated electric energy systems curriculum consisting of *power electronics, power systems*, and *electric machines and drives*. This textbook follows a top-down systems-level approach to *power electronics* to *highlight* interrelationships between these sub-fields within this curriculum, and is intended to cover both the fundamentals and practical design in a single-semester course.

This textbook follows a building-block approach to power electronics that allows an in-depth discussion of several important topics that are left out in a conventional course, for example, designing feedback control, power-factor-correction circuits, soft-switching, and space-vector PWM, which is a PWM technique, far superior to sine-PWM, to name a few. Topics in this book are carefully sequenced to maintain continuity and student interest throughout the course.

In a fast-paced course with proper student background, this book can be covered from front-to-back in one semester. However, the material is arranged in such a way that an instructor, to accommodate the students' background, can either omit an entire topic or cover it quickly to provide just an overview using the accompanying slides, without interrupting the flow.

ACKNOWLEDGMENT

The authors are greatly indebted to two grants to the University of Minnesota from the Office of Naval Research (ONR): N00014-15-1-2391 "Web-Enabled, Instructor-Taught Online Courses," and N00014-19-1-2018 "Developing WBG-Based, Extremely Low-Cost Laboratories for Power Electronics, Motor Drives, and Power System Protection and Relays for National Dissemination." These grants allowed the development of the Workbench simulation platform, which is available free-of-cost for educational purposes. These grants also allowed the development of a low-cost hardware laboratory, available from Sciamble (https://sciamble.com) - a University of Minnesota startup.

The authors would like to sincerely thank Dr. Madhukar Rao Airineni and Dr. Saurabh Tewari for their assistance in developing the LTspice example files and reviewing the material.

ABOUT THE COMPANION WEBSITE

This book is accompanied by companion website:
    www.wiley.com/go/mohan/powerelectronics2e

This website includes

- Solution
- Slides
- Simulation
- CUSP website is a link to another site
- Lab manual is a link to another site

POWER ELECTRONICS:
AN ENABLING TECHNOLOGY

Power electronic systems are essential for energy sustainability, which can be defined as meeting our present needs without compromising the ability of future generations to meet their needs. Using renewable energy for generating electricity and increasing the efficiency of transmitting and consuming it are the twin pillars of sustainability. Some of the applications of power electronics in doing so are as mentioned below:

- Harnessing renewable energy such as wind energy and solar energy using photovoltaics.
- Storage of electricity in batteries and flywheels to offset the variability in the electricity generated by renewables.
- Increasing the efficiency of transmitting electricity.
- Increasing efficiency in consuming the electricity in motor-driven systems and lighting, for example.

This introductory chapter highlights all the points mentioned above, which are discussed in further detail in the context of describing the fundamentals of power electronics in the subsequent chapters.

1.1 INTRODUCTION TO POWER ELECTRONICS

Power electronics is an enabling technology, providing the needed interface between an electrical source and an electrical load, as depicted in Figure 1.1 [1]. The electrical source and the electrical load can, and often do, differ in frequency, voltage amplitudes, and the number of phases. The power electronics interface facilitates the transfer of power from the source to the load by converting voltages and currents from one form to another, in which it is possible for the source and load to reverse roles. The controller shown in Figure 1.1 allows management of the power transfer process in which the conversion of voltages and currents should be achieved with as high energy

Power Electronics A First Course: Simulations and Laboratory Implementations, Second Edition.
Ned Mohan and Siddharth Raju.
© 2023 John Wiley & Sons, Inc. Published 2023 by John Wiley & Sons, Inc.
Companion Website: www.wiley.com/go/mohan/powerelectronics2e

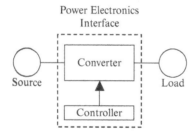

FIGURE 1.1   Power electronics interface between the source and load.

efficiency and high power density as possible. Adjustable-speed electric drives, for example in wind turbines, represent an important application of power electronics.

1.2   APPLICATIONS AND THE ROLE OF POWER ELECTRONICS

Power electronics and drives encompass a wide array of applications. A few important applications and their role are described below.

1.2.1   Powering the Information Technology

Most of the consumer electronics equipment such as personal computers (PCs) and entertainment systems supplied from the utility need very low DC voltages internally. They, therefore, require power electronics in the form of switch-mode DC power supplies for converting the input line voltage into a regulated low DC voltage, as shown in Figure 1.2a. Figure 1.2b shows the distributed architecture typically used in computers in which the incoming AC voltage from the utility is converted into DC voltage, for example, at 24 V. This semi-regulated voltage is distributed within the computer where onboard power supplies in logic-level printed circuit boards convert this 24 V DC input voltage to a lower voltage, for example, 5 V DC, which is very tightly regulated. Very large-scale integration and higher logic circuitry speed require operating voltages much lower than 5 V; hence 3.3 V, 1 V, and eventually, 0.5 V levels would be needed.

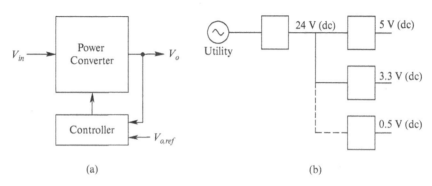

(a)   (b)

FIGURE 1.2   Regulated low-voltage DC power supplies.

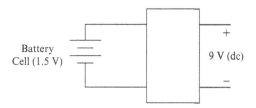

FIGURE 1.3   Boost DC-DC converter needed in cell-operated equipment.

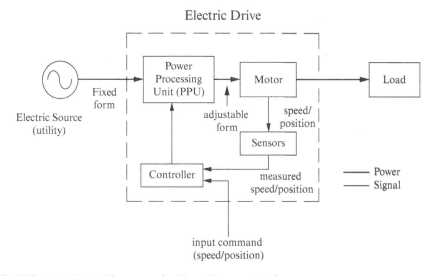

FIGURE 1.4   Block diagram of adjustable-speed drives.

Many devices such as cell phones operate from low battery voltages with one or two battery cells as inputs. However, the electronic circuitry within them requires higher voltages, thus necessitating a circuit to boost input DC to a higher DC voltage as shown in the block diagram of Figure 1.3.

1.2.2   Robotics and Flexible Production

Robotics and flexible production are now essential to industrial competitiveness in a global economy. These applications require adjustable-speed drives for precise speed and position control. Figure 1.4 shows the block diagram of adjustable-speed drives in which the AC input from a 1-phase or a 3-phase utility source is at the line frequency of 50 or 60 Hz. The role of the power electronics interface, as a power-processing unit, is to provide the required voltage to the motor. In the case of a DC motor, DC voltage is supplied with an adjustable magnitude that controls the motor speed. In the case of an AC motor, the power electronics interface provides sinusoidal AC voltages with adjustable amplitude and frequency to control the motor speed. In certain cases, the power electronics interface may be required to allow bidirectional power flow through it, between the utility and the motor load.

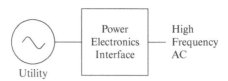

FIGURE 1.5   Power electronics interface required for induction heating.

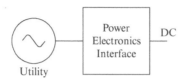

FIGURE 1.6   Power electronics interface required for electric welding.

Induction heating and electric welding, shown in Figures 1.5 and 1.6, respectively, by their block diagrams, are other important industrial applications of power electronics for flexible production.

1.3   ENERGY AND THE ENVIRONMENT: ROLE OF POWER ELECTRONICS IN PROVIDING SUSTAINABLE ELECTRIC ENERGY

As mentioned in the preface of this textbook, power electronics is an enabling technology in providing sustainable electric energy. Most scientists now believe that carbon-based fuels for energy production contribute to climate change, which is threatening human civilization. In the United States, the Department of Energy reports that approximately 40% of all the energy consumed is first converted into electricity. Potentially, the use of electric and plug-in hybrid cars, high-speed rails, and so on, may increase this to even 60%. Therefore, it is essential that we generate electricity from renewable sources such as wind and solar, which, at present, represent only slightly over 4%, build the next-generation smarter grid to utilize renewable resources often in remote locations, and use electricity in more energy-efficient ways. Undoubtedly, using electricity efficiently and generating it from renewable sources are the twin pillars of sustainability, and power electronic systems discussed in this textbook are a key to them both!

1.3.1   Energy Conservation

It's an old adage: a penny saved is a penny earned. Not only does energy conservation lead to financial savings, but it also helps the environment. The pie chart in Figure 1.7 shows the percentages of electricity usage in the United States for various applications. The potential for energy conservation in these applications are discussed below.

1.3.1.1   Electric-Motor Driven Systems
Figure 1.7 shows that electric motors, including their applications in heating, ventilating, and air conditioning (HVAC), are responsible for consuming one-half to

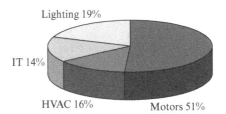

FIGURE 1.7 Percentage use of electricity in various sectors in the US.

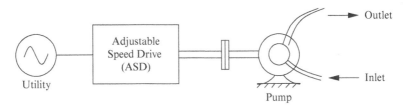

FIGURE 1.8 Role of adjustable-speed drives in pump-driven systems.

two-thirds of all the electricity generated. Traditionally, motor-driven systems run at a nearly constant speed, and their output, for example, the flow rate in a pump, is controlled by wasting a portion of the input energy across a throttling valve. This waste is eliminated by an adjustable-speed electric drive, as shown in Figure 1.8, by efficiently controlling the motor speed, hence the pump speed, by means of power electronics [2].

One out of three new homes in the United States now uses an electric heat pump, in which an adjustable-speed drive can reduce energy consumption by as much as 30% [3] by eliminating on-off cycling of the compressor and running the heat pump at a speed that matches the thermal load of the building. The same is true for air conditioners.

A Department of Energy report [4] estimates that operating all these motor-driven systems more efficiently in the United States could annually save electricity equivalent to the annual electricity usage by the entire state of New York!

1.3.1.2 Lighting Using LEDs
As shown in the pie chart in Figure 1.7, approximately one-fifth of the electricity produced is used for lighting. LEDs (light-emitting diodes) can improve this efficiency by more than a factor of six. They offer a longer lifetime and have become equally affordable as incandescent lamps. They require a power-electronic interface, as shown in Figure 1.9, to convert the line-frequency to supply DC current to the LEDs.

1.3.1.3 Transportation
Electric drives offer huge potential for energy conservation in transportation. While efforts to introduce commercially viable electric vehicles (EVs) continue with progress in battery [5] and fuel cell technologies [6] being reported, hybrid electric vehicles (HEVs) are sure to make a huge impact [7]. According to the US Environmental Protection Agency, the estimated gas mileage of the hybrid-electrical vehicle shown in Figure 1.10 in combined city and highway driving is 48 miles per gallon [8]. This is in

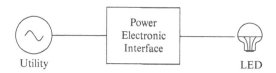

FIGURE 1.9   Power electronics interface required for LED.

FIGURE 1.10   Hybrid electric vehicles with much higher gas mileage.

comparison to the gas mileage of 22.1 miles per gallon for an average passenger car in the United States [9]. Since automobiles are estimated to account for about 20% of the emission of all CO_2 [10], which is a greenhouse gas, doubling the gas mileage of automobiles would have an enormous positive impact.

Conventional automobiles need power electronics for various applications [11]. EVs and HEVs, of course, need power electronics in the form of adjustable-speed electric drives. Add to automobiles other transportation systems, such as light rail, fly-by-wire planes, all-electric ships, and drive-by-wire automobiles, and the conclusion is clear: transportation represents a major application area of power electronics.

1.3.2   Renewable Energy

Clean and renewable energy can be derived from the sun and the wind. In photovoltaic systems, solar cells produce DC, with an I-V characteristic shown in Figure 1.11a that requires a power electronics interface to transfer power to the utility system, as shown in Figure 1.11b.

Wind is the fastest-growing energy resource with enormous potential [12]. Figure 1.12 shows the need for power electronics in wind-electric systems to interface variable-frequency AC to the line-frequency AC voltages of the utility grid.

1.3.3   Utility Applications of Power Electronics

Applications of power electronics and electric drives in power systems are growing rapidly. In distributed generation, power electronics is needed to interface nonconventional energy sources such as wind, photovoltaic, and fuel cells to the utility grid. The

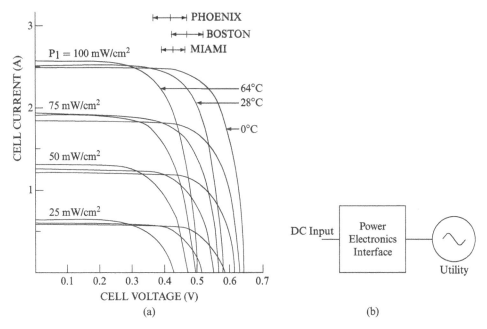

FIGURE 1.11 Photovoltaic systems.

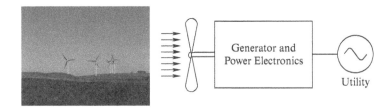

FIGURE 1.12 Wind-electric systems.

use of power electronics allows control over the flow of power on transmission lines, an attribute that is especially significant in a deregulated utility environment. Also, the security and the efficiency aspects of power systems operation necessitate increased use of power electronics in utility applications.

Uninterruptible power supplies (UPS) are used for critical loads that must not be interrupted during power outages. The power electronics interface for UPS, shown in Figure 1.13, has line-frequency voltages at both ends, although the number of phases may be different, and a means for energy storage is usually provided by batteries, which supply power to the load during the utility outage.

1.3.4 Strategic Space and Defense Applications

Power electronics is essential for space exploration and for interplanetary travel. Defense has always been an important application, but it has become critical in the post-September 11th world. Power electronics will play a huge role in tanks, ships, and

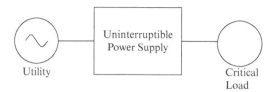

FIGURE 1.13   Uninterruptible power supply (UPS) system.

planes in which replacement of hydraulic drives by electric drives can offer significant cost, weight, and reliability advantages.

1.4   NEED FOR HIGH EFFICIENCY AND HIGH POWER DENSITY

Power electronic systems must be energy-efficient and reliable, have a high power density, thus reducing their size and weight, and be low cost to make the overall system economically feasible. High energy efficiency is important for several reasons: it lowers operating costs by avoiding the cost of wasted energy, contributes less to global warming, and reduces the need for cooling (by heat sinks, discussed later in this book), therefore increasing power density.

We can easily show the relationship in a power electronic system between the energy efficiency, η, and the power density. The energy efficiency of a system in Figure 1.14a is defined in Equation 1.1 in terms of the output power P_o and the power loss P_{loss} within the system as:

$$\eta = \frac{P_o}{P_o + P_{loss}} \tag{1.1}$$

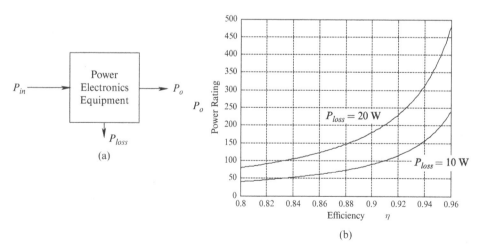

FIGURE 1.14   Power output capability as a function of efficiency.

Equation 1.1 can be rewritten for the output power in terms of the efficiency and the power loss as:

$$P_o = \frac{\eta}{1-\eta} P_{loss} \qquad (1.2)$$

Using Equation 1.2, the output power rating is plotted in Figure 1.14b, as a function of efficiency, for two values of P_{loss}.

In power electronics equipment, the cooling system is designed to transfer dissipated power, as heat, without allowing the internal temperatures to exceed certain limits. Therefore, for an equipment package designed to handle certain power loss dissipation, the plots in Figure 1.14b based on Equation 1.2 show that increasing the conversion efficiency from 84% to 94%, for example, increases the power output capability, same as the power rating, of that equipment by a factor of three. This could mean an increase in the power density, which is the power rating divided by the volume of the package, by approximately the same factor. This is further illustrated by Example 1.1 on the following page.

Example 1.1
A power electronics package is designed to handle 200 W of power dissipation. Compare the two values of the output power capability if the conversion efficiency is increased from 89% to 94%.

Solution   In this example, $P_{loss} = 200\,\text{W}$. Using Equation 1.2, at $\eta = 89\%$, $P_o = \frac{\eta}{1-\eta}$ $P_{loss} = \frac{0.89}{1-0.89} \times 200 \simeq 1.6\,\text{kW}$, and at $\eta = 94\%$, $P_o = \frac{0.94}{1-0.94} \times 200 \simeq 3.13\,\text{kW}$.

This example shows the importance of high energy conversion efficiency, where the power output capability and the power density (in watts per unit volume) of this package are nearly doubled by increasing the efficiency from 89% to 94%.

1.5   STRUCTURE OF POWER ELECTRONICS INTERFACE

By reviewing the role of power electronics in various applications discussed earlier, we can summarize that a power electronics interface is needed to efficiently control the transfer of power between DC-DC, DC-AC, and AC-AC systems. In general, the power is supplied by the utility, and hence, as depicted by the block diagram of Figure 1.15, the line-frequency AC is at one end. At the other end, one of the following is synthesized: adjustable magnitude DC, sinusoidal AC of adjustable frequency and amplitude, or high-frequency AC as in induction heating or systems using high-frequency transformers as an intermediate stage. Applications that do not require utility interconnection can be considered as the subset of the block diagram shown in Figure 1.15.

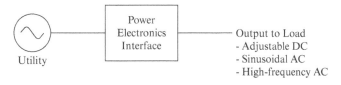

FIGURE 1.15   Block diagram of power electronic interface.

1.5.1 Voltage-Link Structure

To provide the needed functionality to the interface in Figure 1.15, the transistors and diodes, which can block voltage only of one polarity, have led to a commonly used voltage-link-structure, shown in Figure 1.16.

This structure consists of two separate converters, one on the utility side and the other on the load side. The DC ports of these two converters are connected to each other with a parallel capacitor forming a DC-link, across which the voltage polarity does not reverse, thus allowing unipolar voltage-blocking transistors to be used within these converters.

In the structure of Figure 1.16, the capacitor in parallel with the two converters forms a DC voltage-link. Hence, it is called a *voltage-link* (or a *voltage-source*) structure. This structure is used in a very large power range, from a few tens of watts to several megawatts, even extending to hundreds of megawatts in utility applications. Therefore, we will mainly focus on this voltage-link structure in this book.

1.5.2 Current-Link Structure

At extremely high power levels, usually in utility-related applications, which we will discuss in the last two chapters in this book, it may be advantageous to use a *current-link* (also called *current-source*) structure, where, as shown in Figure 1.17, an inductor in series between the two converters acts as a current-link. These converters generally consist of thyristors, and the current in them, as discussed in Chapter 14, is "commutated" from one AC phase to another by means of the AC line voltages.

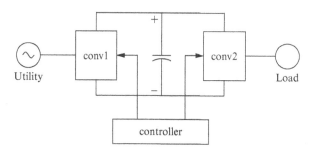

FIGURE 1.16 Voltage-link structure of power electronics interface.

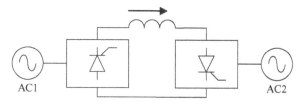

FIGURE 1.17 Current-link structure of power electronics interface.

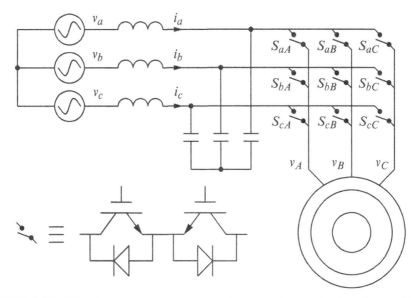

FIGURE 1.18 Matrix converter structure of power electronics interface. [13] / U.S Department of Energy / Public Domain.

1.5.3 Matrix Converters (Direct-Link Structure) [13]

Lately, in certain applications, a *matrix converter* structure, as shown in Figure 1.18 is being reevaluated, where theoretically, there is no energy storage element between the input and the output sides. Therefore, we can consider it to be a direct-link structure where input ports are connected to output ports by switches that can carry currents in both directions when *on* and block voltages of either polarity when *off*. A detailed discussion of matrix converters and their controls can be found in [13].

1.6 VOLTAGE-LINK-STRUCTURE

In the voltage-link structure shown in Figure 1.16 and repeated in Figure 1.19, the role of the utility-side converter is to convert line-frequency utility voltages to an unregulated DC voltage. This can be done by a diode-rectifier circuit such as that discussed in basic electronics courses and also discussed in Chapter 5 of this textbook. However, the power quality concerns often lead to a different structure, discussed in Chapter 6. At

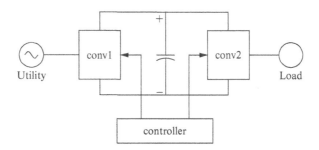

FIGURE 1.19 Load-side converter in a voltage-source structure.

present, we will focus our attention on the load-side converter in the voltage-link structure, where a DC voltage is applied as the input on one end, as shown in Figure 1.19.

Applications dictate the functionality needed of the load-side converter. Based on the desired output of the converter, we can group these functionalities as follows:

Group 1

- Adjustable DC or a low-frequency sinusoidal AC output in
  - DC and AC motor drives
  - uninterruptible power supplies
  - regulated DC power supplies without electrical isolation
  - utility-related applications

Group 2

- High-frequency AC in
  - systems using high-frequency transformers as an intermediate stage
  - induction heating
  - regulated DC power supplies where the DC output voltage needs to be electrically isolated from the input, and the load-side converter internally produces high-frequency AC, which is passed through a high-frequency transformer and then rectified into DC.

We will discuss converters used in applications belonging to both groups. However, we will begin with converters for group-1 applications where the load-side voltages are DC or low-frequency AC.

1.6.1   Switch-Mode Conversion: Switching Power-Pole as the Building Block

Achieving high energy efficiency for applications belonging to either group mentioned above requires switch-mode conversion, where in contrast to linear power electronics, transistors (and diodes) are operated as switches, either on or off.

This switch-mode conversion can be explained by its basic building block, a switching power-pole A, as shown in Figure 1.20a. It effectively consists of a bi-positional switch, which forms a two-port: (1) a voltage-port across a capacitor with a voltage V_{in} that cannot change instantaneously, and (2) a current-port due to the series inductor through which the current cannot change instantaneously. For now, we will assume the switch is ideal with two positions: up or down, dictated by a switching

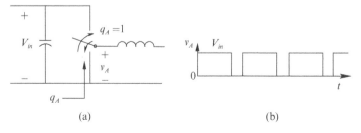

FIGURE 1.20   Switching power-pole as the building block in converters.

signal q_A, which takes on two values: 1 and 0, respectively. The practical aspects of implementing this bi-positional switch are what we will consider in the next chapter.

The bi-positional switch "chops" the input DC voltage V_{in} into a train of high-frequency voltage pulses, shown by v_A waveform in Figure 1.20b, by switching up or down at a high repetition rate, called the switching frequency f_s. Controlling the pulse width within a switching cycle allows control over the switching-cycle-averaged value of the pulsed output, and this pulse-width modulation forms the basis of synthesizing adjustable DC and low-frequency sinusoidal AC outputs, as described in the next section. High-frequency pulses are clearly needed in applications such as compact fluorescent lamps and induction heating and internally in DC power supplies where electrical isolation is achieved by means of a high-frequency transformer. A switch-mode converter consists of one or more such switching power-poles.

1.6.2 Pulse-Width Modulation (PWM) of the Switching Power-Pole (Constant f_s)

For the applications in group 1, the objective of the switching power-pole redrawn in Figure 1.21a is to synthesize the output voltage such that its *switching-cycle average* is of the desired value: DC or AC that varies sinusoidally at a low frequency, compared to f_s. Switching at a constant switching frequency f_s produces a train of voltage pulses in Figure 1.21b that repeat with a constant switching time period T_s, equal to $1/f_s$.

Within each switching cycle with the time period T_s ($= 1/f_s$) in Figure 1.21b, the switching-cycle-averaged value $\bar{v}_A$ of the waveform is controlled by the pulse width T_{up} (during which the switch is in the up position and v_A equals V_{in}), as a ratio of T_s:

$$\bar{v}_A = \frac{T_{up}}{T_s}V_{in} = d_A V_{in}, \quad 0 \le d_A \le 1, \tag{1.3}$$

where $d_A (= T_{up}/T_s)$, which is the average of the q_A waveform as shown in Figure 1.21b, is defined as the duty ratio of the switching power pole A, and the switching-cycle-averaged voltage is indicated by a "–" on top (the overbar symbol). The switching-cycle-averaged voltage and the switch duty ratio are expressed by lowercase letters since they may vary as functions of time. The control over the switching-cycle-averaged value of the output voltage is achieved by adjusting or modulating the pulse width, which later on will be referred to as pulse-width-modulation (PWM). This switching

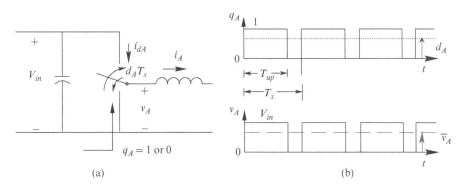

FIGURE 1.21 PWM of the switching power-pole.

power-pole and the control of its output by PWM set the stage for switch-mode conversion with high energy efficiency.

We should note that $\bar{v}_A$ and d_A in the above discussion are discrete quantities, and their values, calculated over a *k-th* switching cycle, for example, can be expressed as $\bar{v}_{A,k}$ and $d_{A,k}$. However, in practical applications, the pulse-width T_{up} changes very slowly over many switching cycles, and hence we can consider them analog quantities as $\bar{v}_A(t)$ and $d_A(t)$ that are continuous functions of time. For simplicity, we may not show their time dependence explicitly.

1.6.3   Switching Power-Pole in a Buck DC-DC Converter: An Example

As an example, we will consider the switching power-pole in a *buck* converter to step down the input DC voltage V_{in}, as shown in Figure 1.22a, where a capacitor is placed in parallel with the load to form a low-pass L-C filter with the inductor, to provide a smooth voltage to the load.

In steady state, the DC (average) input to this L-C filter has no attenuation. Hence, the average output voltage V_o equals the switching-cycle average, $\bar{v}_A$, of the applied input voltage. Based on Equation 1.3, by controlling d_A, the output voltage can be controlled in a range from V_{in} down to 0:

$$V_o = \bar{v}_A = d_A V_{in} \quad (0 \leq V_o \leq V_{in}) \tag{1.4}$$

In spite of the pulsating nature of the instantaneous output voltage $\bar{v}_A(t)$, the series inductance at the current-port of the pole ensures that the current $i_L(t)$ remains relatively smooth, as shown in Figure 1.22b.

Example 1.2

In the converter of Figure 1.22a, the input voltage $V_{in} = 20$ V. The output voltage $V_o = 12$ V. Calculate the duty ratio d_A and the pulse width T_{up}, if the switching frequency $f_s = 200$ kHz.

Solution $\bar{v}_A = V_o = 12$ V. Using Equation 1.4, $d_A = \frac{V_o}{V_{in}} = \frac{12}{20} = 0.6$ and $T_s = \frac{1}{f_s} = 5\,\mu s$. Therefore, as shown in Figure 1.23, $T_{up} = d_A T_s = 0.6 \times 5\,\mu s$.

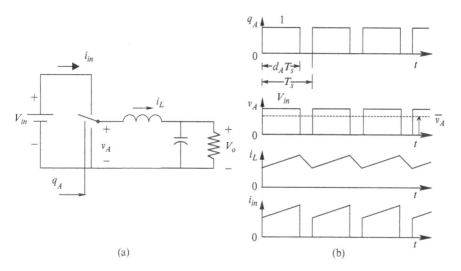

(a)   (b)

FIGURE 1.22   Switching power-pole in a buck converter.

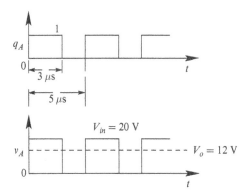

FIGURE 1.23 Waveforms in the converter of Example 1.2.

1.6.3.1 Realizing the Bi-Positional Switch in a Buck Converter

As shown in Figure 1.24a, the bi-positional switch in the power pole can be realized by using a transistor and a diode. When the transistor is gated on (through a gate circuitry, discussed in the next chapter, by a switching signal $q_A = 1$), it carries the inductor current, and the diode is reversed biased, as shown in Figure 1.24b. When the transistor is switched off ($q_A = 0$), as shown in Figure 1.24c, the inductor current "freewheels" through the diode until the next switching cycle when the transistor is turned back on. The switching waveforms, shown earlier in Figure 1.22b, are discussed in detail in Chapter 3.

In the switch-mode circuit of Figure 1.22a, the higher the switching frequency, that is, the frequency of the pulses in the $v_A(t)$ waveform, the smaller the values needed for the low-pass L-C filter. On the other hand, higher switching frequency results in higher switching losses in the bi-positional switch, which is the subject of the next chapter. Therefore, an appropriate switching frequency must be selected, keeping these trade-offs in mind.

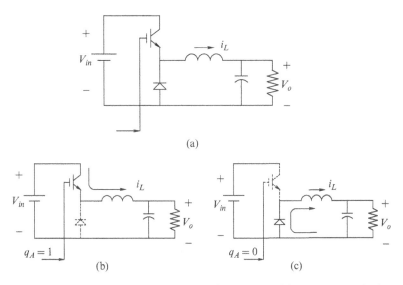

FIGURE 1.24 Transistor and diode forming a switching power-pole in a buck converter.

1.7 RECENT ADVANCES IN SOLID-STATE DEVICES BASED ON WIDE BANDGAP (WBG) MATERIALS

There have been significant advances in the wide bandgap (WBG) materials that have made WBG devices such as GaN- and SiC-based diodes and MOSFETs available at reasonable prices for power electronics applications. These include SiC-MOSFETs with voltage-blocking capabilities as high as 1.7 kV.

In comparison to these WBG devices, the silicon power device technology is mature, and only minor evolutionary improvements are likely. These devices are limited to junction temperatures < 150–200 °C. However, power electronics applications are extending to ever higher voltages, higher frequencies, higher temperatures, and higher efficiencies simultaneously. This makes the use of Si-based devices in future applications increasingly difficult and expensive.

WBG devices offer a potential solution where they are especially better for higher voltages (> 1000 V) as well as in low-voltage applications below 600 volts or so. They are faster, have lower losses, and higher operating temperatures to minimize the thermal management systems.

WBG devices operate by the same physical mechanisms as Si devices and have similar geometries. Their better performance is due to superior material properties. The WBG materials of interest are gallium nitride (GaN) and silicon carbide (SiC). The WBG devices are fundamentally superior to Si devices for the basic reason of 10 times larger breakdown field strength of the WBG materials that enable faster switching with lower losses, higher temperature operation, and higher thermal conductivity.

A variety of SiC-devices are commercially available, and SiC MOSFETs and diodes will dominate power electronic applications above 1000 V. Similarly, GaN-based diodes and MOSFETs will dominate power electronic applications below 600 V. Their applications are shown in Figure 1.25 as functions of the operating frequencies and power levels.

Further details on WBGs can be found in [13].

1.8 USE OF SIMULATION AND HARDWARE PROTOTYPING

Throughout this book, modeling tools are used to facilitate discussion and provide an in-depth understanding of the concepts in power electronics. LTspice and Sciamble™ Workbench [15] are computer simulation tools used to demonstrate all topics in this book.

LTspice is a widely used SPICE-based circuit simulator. Sciamble™ Workbench is a mathematical simulation tool developed at the University of Minnesota. All examples and key concepts explained in this book are also simulated using both LTspice and Workbench, and the results are provided on the accompanying website.

A noteworthy motivation for using Sciamble™ Workbench is its seamless transition between mathematical simulation and hardware prototyping modes. Hardware prototyping simplifies the development of a real-time controller enabling rapid laboratory experimentation of concepts covered in this book. Real-world experimentation enables a more in-depth and practical understanding of the contents of this book. Sciamble™'s power electronics hardware kit is used for laboratory implementation of all the topics in this book.

All the simulation files using both LTspice and Sciamble™ Workbench, as well as the manual for laboratory implementation using Workbench, are available on the accompanying website.

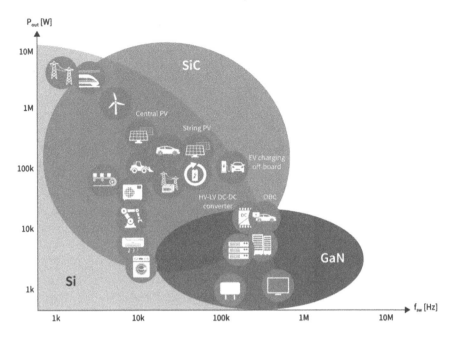

FIGURE 1.25 Applications of SiC- and GaN-based devices.
[*Source*: Graph used under kind permission of Infineon Technologies AG] [14]; https://www.infineon.com/cms/en/product/technology/wide-bandgap-semiconductors-sic-gan.

REFERENCES

1. N. Mohan, T.M. Undeland, and W.P. Robbins, *Power Electronics: Converters, Applications and Design*, 3rd Edition (New York: John Wiley & Sons, 2003).

2. N. Mohan and R.J. Ferraro, "Techniques for Energy Conservation in AC Motor Driven Systems," EPRI-Report EM-2037, Project 1201 1213, September 1981.

3. N. Mohan and J. Ramsey, "Comparative Study of Adjustable-Speed Drives for Heat Pumps," EPRI-Report EM-4704, Project 2033–4, August 1986.

4. *Improving Motor and Drive System Performance: A Sourcebook for Industry, Industrial Technologies Program (ITP)* (Book), US Department of Energy (energy.gov). https://www1.eere.energy.gov/manufacturing/tech_assistance/pdfs/motor.pdf.

5. "Significant Progress in Lithium-air Battery Development," ScienceDaily. https://www.sciencedaily.com/releases/2021/05/210506104801.htm.

6. "Progress in Hydrogen and Fuel Cells." US Department of Energy. https://www.energy.gov/eere/fuelcells/articles/progress-hydrogen-and-fuel-cells.

7. "Alternative Fuels Data Center: Emissions from Electric Vehicles," US Department of Energy. https://afdc.energy.gov/vehicles/electric_emissions.html.

8. "Toyota Electric Vehicles." https://www.toyota.com/electrified.

9. "Alternative Fuels Data Center: Maps and Data—Average Fuel Economy by Major Vehicle Category," US Department of Energy. https://afdc.energy.gov/data/10310.

10. "Car Emissions and Global Warming," Union of Concerned Scientists (Ucsusa.org). https://www.ucsusa.org/resources/car-emissions-global-warming.

11. "Power Electronics in Automotive Applications," Elprocus. https://www.elprocus.com/power-electronics-in-automotive-applications.

12. The American Clean Power Association. cleanpower.org.

13. N. Mohan, W. Robins, T. Undeland, and S. Raju, *Power Electronics for Grid-Integration of Renewables: Analysis, Simulations and Hardware Lab* (New York: John Wiley & Sons, 2023).

14. "Wide Bandgap Semiconductors (SiC/GaN)," Infineon. https://www.infineon.com/cms/en/product/technology/wide-bandgap-semiconductors-sic-gan.

15. Sciamble™ Workbench platform. https://sciamble.com.

PROBLEMS

Applications and Energy Conservation

1.1 A US Department of Energy report [4] estimates that over 122 billion kWh/year can be saved in the manufacturing sector in the United States by using mature and cost-effective conservation technologies. Calculate (a) how many 1,100-MW generating plants are needed to operate constantly to supply this wasted energy, and (b) the annual savings in dollars if the cost of electricity is 0.12 cents/kWh.

1.2 In a process, a blower is used with the flow-rate profile shown in Figure P1.1a.

Using the information in Figure P1.1b, estimate the percentage reduction in power consumption resulting from using an adjustable-speed drive rather than a system with (a) an outlet damper and (b) an inlet vane.

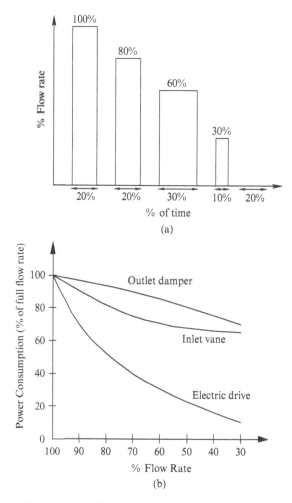

FIGURE P1-1   Flowrate profile.

1.3   In a system, if the system size is based on power dissipation capacity, calculate the improvement in the power density if the efficiency is increased from 87% to 95%.

1.4   The electricity generation in the United States is approximately 3.8×10^9 MW-hrs. Figure 1.7 shows that 16% of that is used for heating, ventilating, and air conditioning. As much as 30% of the energy can be saved in such systems by using adjustable-speed drives. On this basis, calculate the savings in energy per year and relate that to 1,100-MW generating plants needed to operate constantly to supply this wasted energy.

1.5   The total amount of electricity that could potentially be generated from wind in the United States has been estimated at 10.8×10^9 MW-hrs annually. If one-tenth of this potential is developed, estimate the number of 1.5 MW windmills that would be required, assuming that on average a windmill produces only 30% of the energy that it is capable of.

1.6   In Problem 1.5, if each 1.5 MW windmill has all its output power flowing through the power electronics interface, estimate the total rating of these interfaces in MW.

1.7   As the pie chart of Figure 1.7 shows, lighting in the United States consumes 19% of the generated electricity. LEDs consume power one-sixth of that consumed by the incandescent lamps for the same light output. Electricity generation in the United States is approximately 3.8×10^9 MW-hrs. Based on this information, estimate the savings in MW-hrs annually, assuming that all lighting at present is by incandescent lamps, which are to be replaced by LEDs.

1.8   An electric-hybrid vehicle offers 52 miles per gallon in mixed (city and highway) driving conditions according to the US Environmental Protection Agency. This is in comparison to the gas mileage of 22.1 miles per gallon for an average passenger car in the United States, with an average of 11,766 miles driven per year. Calculate the savings in terms of barrels of oil per automobile annually if a conventional car is replaced by an electric-hybrid vehicle. In calculating this, assume that a barrel of oil that contains 42 gallons yields approximately 20 gallons of gasoline.

1.9   We can project that there are 150 million cars in the United States. Using the results of Problem 1.8, calculate the total annual reduction of carbon into the atmosphere if the consumption of each gallon of gasoline releases approximately 5 pounds of carbon.

1.10 Relate the savings of barrels of oil annually, as calculated in Problem 1.9, to the imported oil if the United States imports approximately 35,000 million barrels of crude oil per year.

1.11 Fuel-cell systems that also utilize the heat produced can achieve efficiencies approaching 80%, more than double of the gas-turbine-based electrical generation. Assume that 20 million households produce an average of 5 kW. Calculate the percentage of electricity generated by these fuel-cell systems compared to the annual electricity generation in the United States of 3.8×10^9 MW-hrs.

1.12 Induction cooking based on power electronics is estimated to be 80% efficient compared to 55% for conventional electric cooking. If an average home consumes 2 kW-hrs daily using conventional electric cooking, and that 20 million households switch to induction cooking in the United States, calculate the annual savings in electricity usage.

1.13 Assume the average energy density of sunlight to be 800 W/m^2 and the overall photovoltaic system efficiency to be 10%. Calculate the land area covered with photovoltaic cells needed to produce 1,000 MW, the size of a typical large central power plant.

1.14 In Problem 1.13, the solar cells are distributed on top of roofs, each in an area of 40 m². Calculate the number of homes needed to produce the same power.

1.15 Describe the role of power electronics in harnessing onshore and offshore wind energy.

1.16 Describe the role of power electronics in photovoltaic systems to harness solar energy.

1.17 Describe the role of power electronics in battery storage and fuel-cell systems.

1.18 Describe the role of power electronics in electric vehicles, hybrid-electric vehicles, and plug-in hybrid-electric vehicles.

1.19 Describe the role of power electronics in transportation systems by means of high-speed rails.

1.20 Describe the role of power electronics for energy conservation in lighting using LEDs.

1.21 Describe the role of power electronics for energy conservation in heat pumps and air conditioners.

1.22 Describe the role of power electronics in transmitting electricity by means of high-voltage DC (HVDC) transmission.

PWM of the Switching Power-Pole

1.23 In a buck converter, the input voltage V_{in} = 12 V. The output voltage V_o is required to be 9 V. The switching frequency f_s = 400 kHz. Assuming an ideal switching power-pole, calculate the pulse width T_{up} of the switching signal and the duty ratio d_A of the power pole.

1.24 In the buck converter of Problem 1.23, assume the current through the inductor to be ripple-free with a value of 1.5 A (the ripple in this current is discussed later in Chapter 3). Draw the waveforms of the voltage v_A at the current-port and the input current i_{in} at the voltage-port.

1.25 Using the input-output specifications given in Problem 1.23, calculate the maximum energy efficiency expected of a linear regulator where the excess input voltage is dropped across a transistor, that functions as a controllable resistor and is placed in series between the input and the output.

Simulation Problems

1.26 In the circuit diagram of the buck converter in Figure 1.22a, the low-pass filter has the following values: L = 5 μH and C = 100 μF. The output load resistance is R = 0.5 Ω. The input voltage v_A is a pulse waveform between 0 and 10 volts, with a pulse width of 0.75 μs and a frequency f_s = 100 kHz.

   (a) Plot the input voltage v_A and the output voltage v_o for the last 10 switching cycles after v_o waveform has reached its steady state.

   (b) How does V_o relate to $\bar{v}_A$(the average of v_A)?

   (c) What is the ratio of the switching frequency to the L-C resonance frequency? By means of Fourier analysis, compute the attenuation of the fundamental-frequency component in the input voltage by the filter at the switching frequency?

1.27 In Problem 1.26, plot the gain of the transfer function $V_o(s)/V_A(s)$ in dB, as a function of frequency. Does the frequency at which the transfer-function gain is peaking coincide with the L-C resonance frequency? Calculate the attenuation of the fundamental-frequency component by the filter. How does it compare with that obtained by the Fourier analysis in Problem 1.26c?

2

DESIGN OF SWITCHING POWER-POLES

In the previous chapter, we discussed the role of power electronics in energy sustainability. Power electronics is an applied field of study where theory must be translated into practice to realize the immediate challenges that we face. We also discussed a switching power-pole as the building block of power electronic converters, consisting of an ideal bi-positional switch with an ideal transistor and an ideal diode and pulse-width modulation (PWM) to control the output. In this chapter, we will discuss the availability of various power semiconductor devices that are essential in power electronic systems, their switching characteristics, and various trade-offs in designing a switching power-pole, which can be used in various applications discussed in Chapter 1. We will also briefly discuss a PWM-IC, which is used in regulating and controlling the average output of such switching power-poles.

2.1 POWER TRANSISTORS AND POWER DIODES [1]

The power-level diodes and transistors have evolved over decades from their signal-level counterparts to the extent that they can handle voltages and currents in kilovolts and kiloamperes, respectively, with fast switching times of the order of a few tens of ns to a few μs. Moreover, these devices can be connected in series and parallel to satisfy any voltage and current requirements.

The selection of power diodes and power transistors in a given application is based on the following characteristics:

1. *Voltage rating*: The maximum instantaneous voltage that the device is required to block in its off state, beyond which it breaks down, that is, irreversible damage occurs.
2. *Current rating*: The maximum current, expressed as instantaneous, average, and/or RMS, that a device can carry in its on state, beyond which excessive heating within the device destroys it.
3. *Switching speeds*: The speeds with which a device can make a transition from its on state to off state, or vice versa. Small switching times associated with fast-switching

Power Electronics A First Course: Simulations and Laboratory Implementations, Second Edition.
Ned Mohan and Siddharth Raju.
© 2023 John Wiley & Sons, Inc. Published 2023 by John Wiley & Sons, Inc.
Companion Website: www.wiley.com/go/mohan/powerelectronics2e

devices result in low switching losses, or considering it differently, fast-switching devices can be operated at high switching frequencies with acceptable switching power losses.

4. *On-state voltage*: The voltage drop across a device during its on state while conducting a current. The smaller this voltage, the smaller will be the on-state power loss.

2.2   SELECTION OF POWER TRANSISTORS [2–5]

Transistors are controllable switches, which are available in several forms for switch-mode power electronics applications:

- MOSFETs (metal-oxide-semiconductor field-effect transistors)
- IGBTs (insulated-gate bipolar transistors)
- IGCTs (integrated gate-commutated thyristors)
- GTOs (gate turn-off thyristors)
- Niche devices for power electronics applications, such as BJTs (bipolar junction transistors), SITs (static induction transistors), MCTs (MOS-controlled thyristors), and so on.

In switch-mode converters, there are two types of transistors that are primarily used: MOSFETs are typically used below a few hundred volts at switching frequencies in excess of 100 kHz, whereas IGBTs dominate very large voltage, current, and power ranges extending to MW levels, provided the switching frequencies are below a few tens of kHz. IGCTs and GTOs are used in utility applications of power electronics at power levels beyond a few MWs. Figure 1.25 shows the capabilities and applications of SiC- and GaN-based devices.

The following subsections provide a brief overview of MOSFET and IGBT characteristics and capabilities.

2.2.1   MOSFETs

In applications at voltages below 200 V and switching frequencies in excess of 100 kHz, MOSFETs are clearly the device of choice because of their low on-state losses in low voltage ratings, their fast switching speeds, and a high-impedance gate that requires a small voltage and charge to facilitate the on/off transition. The circuit symbol of an n-channel MOSFET is shown in Figure 2.1a. It consists of three terminals: drain (D), source (S), and gate (G). The forward current in a MOSFET flows from the drain to the source terminal. MOSFETs can block only the forward polarity voltage; that is, a positive v_{DS}. They cannot block a negative polarity voltage due to an intrinsic antiparallel diode, which can be used effectively in most switch-mode converter designs. MOSFET I-V characteristics for various gate voltage values are shown in Figure 2.1b.

For gate voltages below a threshold value $v_{GS(th)}$, typically in the range of 2 to 4 V, a MOSFET is completely off, as shown by the I-V characteristics in Figure 2.1b, and approximates an open switch. Beyond $v_{GS(th)}$, the drain current i_D through the MOSFET depends on the applied gate voltage v_{GS}, as shown by the transfer characteristic shown

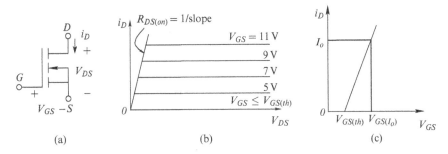

FIGURE 2.1   MOSFET: (a) symbol, (b) I-V characteristics, (c) transfer characteristic.

in Figure 2.1c, which is valid almost for any value of the voltage v_{DS} across the MOSFET (note the horizontal nature of I-V characteristics in Figure 2.1b). To carry $i_D(=I_o)$ would require a gate voltage of a value at least equal to $v_{GS(I_o)}$, as shown in Figure 2.1c. Typically, a higher gate voltage, approximately 10 V, is maintained in order to keep the MOSFET in its on state and carrying $i_D(=I_o)$.

In its on state, a MOSFET approximates a very small resistor $R_{DS(on)}$, and the drain current that flows through it depends on the external circuit in which it is connected. The on-state resistance, the inverse of the slope of I-V characteristics as shown in Figure 2.1b, is a strong function of the blocking voltage rating V_{DSS} of the device:

$$R_{DS(on)} \alpha V_{DSS}^{2.5 \, to \, 2.7} \tag{2.1}$$

The relationship in Equation 2.1 explains why MOSFETs in low-voltage applications, at less than 200 V, are an excellent choice. The on-state resistance goes up with the junction temperature within the device, and proper heatsinking must be provided to keep the temperature below the design limit.

2.2.2   IGBTs

IGBTs combine ease of control, as in MOSFETs, with low on-state losses even at fairly high voltage ratings. Their switching speeds are sufficiently fast for switching frequencies up to 30 kHz. Therefore, they are used for converters in a vast voltage and power range—from a fractional kilowatt to several megawatts where switching frequencies required are below a few tens of kilohertz.

The circuit symbol for an IGBT is shown in Figure 2.2a, and the I-V characteristics are shown in Figure 2.2b. Similar to MOSFETs, IGBTs have a high impedance gate, which requires only a small amount of energy to switch the device. IGBTs have a small on-state voltage, even in devices with a large blocking-voltage rating, for example, V_{on} is approximately 2 V in 1200-V devices.

IGBTs can be designed to block negative voltages, but most commercially available devices, by design to improve other properties, cannot block any appreciable reverse polarity voltage. IGBTs have turn-on and turn-off times on the order of a microsecond and are available as modules in ratings as large as 3.3 kV and 1200 A. Voltage ratings of up to 5 kV are projected.

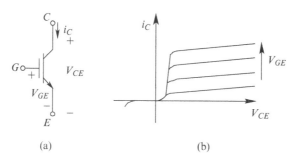

(a)   (b)

FIGURE 2.2   IGBT: (a) symbol, (b) I-V characteristics.

2.2.3   Power-Integrated Modules and Intelligent-Power Modules [2–4]

Power electronic converters, for example, for three-phase AC drives, require six power transistors. Power-integrated modules (PIMs) combine several transistors and diodes in a single package. Intelligent-power modules (IPMs) include power transistors with the gate-drive circuitry. The input to this gate-drive circuitry is a signal that comes from a microprocessor or an analog integrated circuit, and the output drives the MOSFET gate. IPMs also include fault protection and diagnostics, thereby immensely simplifying the power electronics converter design.

2.2.4   Costs of MOSFETs and IGBTs

As these devices evolve, their relative costs continue to decline. The costs of single devices are approximately 0.25 cents/A for 600-V devices and 0.50 cents/A for 1200-V devices. Power modules for the 3 kV class of devices cost approximately 1 cent/A.

2.3   SELECTION OF POWER DIODES

The circuit symbol of a diode is shown in Figure 2.3a, and its I-V characteristic in Figure 2.3b shows that a diode is an uncontrolled device that blocks reverse polarity voltage. Power diodes are available in large ranges of reverse voltage blocking and forward current carrying capabilities. Similar to transistors, power diodes are available in several forms as follows, and their selection must be based on their application:

- Line-frequency diodes
- Fast-recovery diodes
- Schottky diodes
- SiC-based Schottky diodes

Rectification of line-frequency AC to DC can be accomplished by slower switching *p-n* junction power (line-frequency) diodes, which have relatively a slightly lower on-state voltage drop and are available in voltage ratings of up to 9 kV and current ratings of up to 5 kA. The on-state voltage drop across these diodes is usually on the order of 1 to 3 V, depending on the voltage blocking capability.

Switch-mode converters operating at high switching frequencies, from several tens of kilohertz to several hundred kilohertz, require fast-switching diodes. In

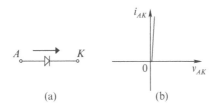

FIGURE 2.3 Diode: (a) symbol, (b) I-V characteristic.

such applications, fast-recovery diodes, also formed by *p-n* junction as the line-frequency diodes, must be selected to minimize switching losses associated with the diodes.

In applications with very low output voltages, the forward voltage drop of approximately a volt across the conventional *p-n* junction diodes becomes unacceptably high. In such applications, another type of device called the Schottky diode is used with a voltage drop in the range of 0.3 to 0.5 V. Being majority-carrier devices, Schottky diodes switch extremely fast and keep switching losses to a minimum.

All devices, transistors, and diodes discussed above are silicon-based. Lately, silicon-carbide (SiC)–based Schottky diodes in voltage ratings of up to 1200 V have become available [6]. In spite of their large on-state voltage drop (1.7 V, for example), their capability to switch with a minimum of switching losses makes them attractive in converters with voltages in excess of a few hundred volts.

2.4 SWITCHING CHARACTERISTICS AND POWER LOSSES IN POWER POLES

In switch-mode converters, it is important to understand the switching characteristics of the switching power-pole. As discussed in the previous chapter, the power pole in a buck converter shown in Figure 2.4a is implemented using a transistor and a diode, where the current through the current port is assumed to be a constant DC, I_o, for discussing switching characteristics. We will assume the transistor to be a MOSFET, although a similar discussion applies if an IGBT is selected.

For the *n*-channel MOSFET (*p*-channel MOSFETs have poor characteristics and are seldom used in power applications), the typical i_D versus v_{DS} characteristics are shown in Figure 2.4b for various gate voltages. The off-state and the on-state operating points are labeled on these characteristics.

Switching characteristics of MOSFETs, and hence of the switching power-pole in Figure 2.4a, are dictated by a combination of factors: the speed of charging and discharging of junction capacitances within the MOSFET, i_D versus v_{DS} characteristics, the circuit of Figure 2.4a in which the MOSFET is connected, and its gate-drive circuitry. We should note that the MOSFET junction capacitances are highly nonlinear functions of drain-source voltage and go up dramatically by several orders of magnitude at lower voltages. In Figure 2.4a, the gate-drive voltage for the MOSFET is represented as a voltage source, which changes from 0 to V_{GG}, approximately 10 V, and vice versa, and charges or discharges the gate through a resistance R_{gate} that is the sum of external resistance R_{GG} and the internal gate resistance.

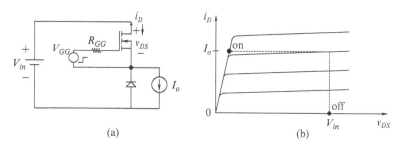

FIGURE 2.4   MOSFET in a switching power-pole.

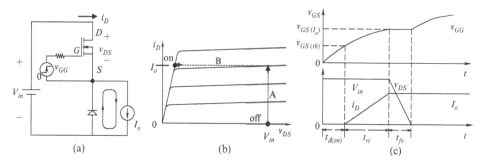

FIGURE 2.5   MOSFET turn-on.

Only a simplified explanation of the turn-on and turn-off characteristics, assuming an ideal diode, is presented here. The switching details, including the diode reverse recovery, are discussed in the Appendix to this chapter.

2.4.1   Turn-On Characteristic

Prior to turn-on, Figure 2.5a shows the circuit in which the MOSFET is off and blocks the input DC voltage V_{in}; I_o freewheels through the diode. By applying a positive voltage to the gate, the turn-on characteristic describes how the MOSFET goes from the off-point to the on-point in Figure 2.5b. To turn the MOSFET on, the gate-drive voltage goes up from 0 to V_{GG}, and it takes the gate drive a finite amount of time, called the turn-on delay time $t_{d(on)}$, to charge the gate-source capacitance through the gate-circuit resistance R_{gate} to the threshold value of $V_{GS(th)}$. During this turn-on delay time, the MOSFET remains off and I_o continues to freewheel through the diode, as shown in Figure 2.5a.

For the MOSFET to turn on, first, the current i_D through it rises. As long as the diode is conducting a positive net current ($I_o - i_D$), the voltage across it is zero, and the MOSFET must block the entire input voltage V_{in}. Therefore, during the current rise time t_{ri}, the MOSFET voltage and the current are along the trajectory A in the I-V characteristics of Figure 2.5b and are plotted in Figure 2.5c. Once the MOSFET current reaches I_o, the diode becomes reverse biased, and the MOSFET voltage falls. During the voltage fall time t_{fv}, as depicted by the trajectory B in Figure 2.5b and the plots in Figure 2.5c, the gate-to-source voltage remains at $V_{GS(I_o)}$. Once the turn-on transition is complete, the gate charges to the gate-drive voltage V_{GG}, as shown in Figure 2.5c.

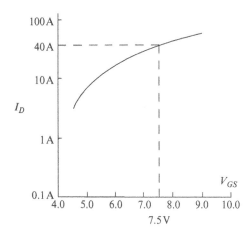

FIGURE 2.6 MOSFET transfer characteristic.

Example 2.1

In the converter of Figure 2.4a, the transistor is a MOSFET that carries a current of 5 A when it is fully on. If the current through the transistor is to be limited to 40 A during a malfunction, in which case the entire input voltage of 50 V appears across the transistor, what should be the maximum on-state gate voltage that the gate-drive circuit should provide? Assume the junction temperature T_j of the MOSFET to be 175 °C.

Solution The transfer characteristic of this MOSFET is shown in Figure 2.6. It shows that if $V_{GS} = 7.5$ V is used, the current through the MOSFET will be limited to 40 A.

2.4.2 Turn-Off Characteristic

The turn-off sequence is the opposite of the turn-on process. Prior to turn-off, the MOSFET is conducting I_o, and the diode is reverse biased, as shown in Figure 2.7a. The MOSFET I-V characteristics are replotted in Figure 2.7b. The turn-off characteristic describes how the MOSFET goes from the on-point to the off-point in Figure 2.7b. To turn the MOSFET off, the gate-drive voltage goes down from V_{GG} to 0. It takes the gate drive a finite amount of time, called the turn-off delay time $t_{d(off)}$, to discharge the gate-source capacitance through the gate-circuit resistance R_{gate}, from a voltage of V_{GG} to $V_{GS(I_o)}$. During this turn-off delay time, the MOSFET remains on.

For the MOSFET to turn off, the output current must be able to freewheel through the diode. This requires the diode to become forward biased, and thus the voltage across the MOSFET must rise, as shown by the trajectory C in Figure 2.7b, while the current through it remains I_o. During this voltage rise time t_{rv}, the voltage and current are plotted in Figure 2.7c, while the gate-to-source voltage remains at $V_{GS(I_o)}$. Once the voltage across the MOSFET reaches V_{in}, the diode becomes forward biased, and the MOSFET current falls. During the current fall time t_{fi}, depicted by the trajectory D in Figure 2.7b and the plots in Figure 2.7c, the gate-to-source voltage declines to $V_{GS(th)}$. Once the turn-off transition is complete, the gate discharges to 0.

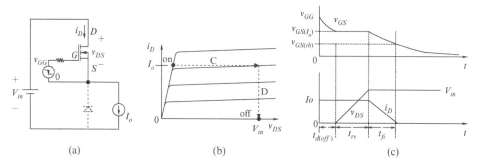

FIGURE 2.7 MOSFET turn-off.

2.4.3 Calculating Power Losses within the MOSFET (Assuming an Ideal Diode)

Power losses in the gate-drive circuitry are negligibly small except at very high switching frequencies. The primary source of power losses is across the drain and the source, which can be divided into two categories: the conduction loss and the switching losses. Both of these are discussed in the following sections.

2.4.3.1 Conduction Loss

In the on-state, the MOSFET conducts a drain current for an interval T_{on} during every switching time period T_s, with the switch duty ratio $d = T_{on} / T_s$. Assuming this current to be at a constant I_o during dT_s, since it is zero during the rest of the switching time period, the RMS value of the MOSFET current is

$$I_T(\text{rms}) = \sqrt{d} I_o. \tag{2.2a}$$

Hence, the average power loss in the on-state resistance $R_{DS(on)}$ of the MOSFET is

$$P_{cond} = R_{DS(on)} d \, I_o^2. \tag{2.2b}$$

As pointed out earlier, $R_{DS(on)}$ varies significantly with the junction temperature, and data sheets often provide its value at the junction temperature equal to 25 °C. However, it will be more realistic to use twice this resistance value, which corresponds to the junction temperature equal to 120 °C, for example. Equation 2.3 can be refined to account for the effect of a ripple in the drain current on the conduction loss. The conduction loss is highest at the maximum load on the converter when the drain current would also be at its maximum.

2.4.3.2 Switching Losses

At high switching frequencies, switching power losses can be even higher than the conduction loss. The switching waveforms for the MOSFET voltage v_{DS} and the current i_D, corresponding to the turn-on and the turn-off trajectories in Figures 2.5 and 2.7, assuming they are linear with time, are replotted in Figure 2.8.

During each transition from on to off, and vice versa, the transistor has simultaneously high voltage and current, as seen from the switching waveforms in Figure 2.8.

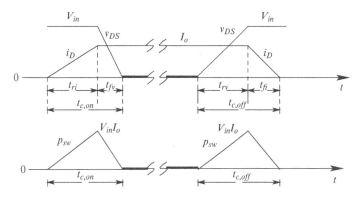

FIGURE 2.8 MOSFET switching losses.

The instantaneous power loss $p_{sw}(t)$ in the transistor is the product of v_{DS} and i_D, as plotted. The average value of the switching losses from the plots in Figure 2.8 is

$$P_{sw} = \frac{1}{2}V_{in}I_o(t_{c,on} + t_{c,off})f_s, \tag{2.3}$$

where $t_{c,on}$ and $t_{c,off}$ are defined in Figure 2.8 as the sum of the rise and the fall times associated with the MOSFET voltage and current:

$$t_{c,on} = t_{ri} + t_{fv} \tag{2.4}$$

$$t_{c,off} = t_{rv} + t_{fi}. \tag{2.5}$$

2.4.4 Gate Driver Integrated Circuits (ICs) with Built-in Fault Protection [2]

Application-specific ICs (ASICs) for controlling the gate voltages of MOSFETs and IGBTs greatly simplify converter design by including various protection features, for example, the over-current protection that turns off the transistor under fault conditions. This functionality, as discussed earlier, is integrated into intelligent-power modules along with the power semiconductor devices.

The IC to drive the MOSFET gate is shown in Figure 2.9 in a block diagram form. To drive the high-side MOSFET in the power pole, the input signal v_c from the controller is referenced to the logic level ground, V_{CC} is the logic supply voltage, and V_{ext} to drive the gate is supplied by an isolated power supply referenced to the MOSFET source S.

One of the ICs for this purpose is UCC21710, which can source up to 10 A for gate charging and sink up to 10 A for gate discharging. It has built-in overcurrent fault protection capability, in which the transistor current, measured via the voltage drop across a small resistance in series with the transistor, disables the gate-drive voltage under over-current conditions.

For extremely low-cost designs, the bootstrap driver shown in Figure 2.10a is a suitable alternative to the more expensive optically or galvanically isolated gate driver.

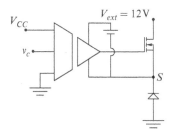

FIGURE 2.9   Gate-driver IC functional diagram.

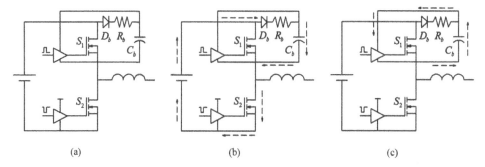

FIGURE 2.10   Bootstrap gate-driver circuit and operation.

These are suitable for converters with a pair of series-connected devices that are driven in a complementary manner.

Unlike an isolated gate driver, which requires an isolated power supply to source the buffer that drives the high-side switch S_1 in Figure 2.10a, the bootstrap driver uses the energy stored in the capacitor C_b to source the buffer. This capacitor is charged up during the normal operation of the converter. When S_1 is off and S_2 is on, the capacitor is charged from the input DC power supply, as shown in Figure 2.10b. The resistor R_b limits the current flowing into the capacitor, and the voltage across the capacitor is clamped using an active circuit typically present within the driver or using an external Zener diode. When S_2 turns off, the diode D_b gets reverse biased. Thus, the voltage across C_b is available for the buffer of S_1 to source the gate turn on. During this interval, the capacitor gets discharged, as shown in Figure 2.10c.

There are a few shortcomings associated with bootstrap drives. The maximum on-time of the high-side device is limited so as to provide enough time for the bootstrap capacitor to charge. It lowers the overall efficiency due to the power loss associated with R_b during every charging cycle, which is more pronounced at high voltages, and limits its use to mostly converters operating within 100 V.

2.5 JUSTIFYING SWITCHES AND DIODES AS IDEAL

Product reliability and high energy efficiency are two very important criteria. In designing converters, it is essential that in the selection of semiconductor devices, we consider their voltage and current ratings with proper safety margins for reliability purposes. Achieving high energy efficiency from converters requires that in selecting power devices, we consider their on-state voltage drops and their switching characteristics. We also need to calculate the heat that needs to be removed for proper thermal design.

Typically, the converter energy efficiency of approximately 90% or higher is realized. This implies that semiconductor devices have a very small loss associated with them. Therefore, in analyzing various converter topologies and comparing them against each other, we can assume transistors and diodes are ideal. Their non-idealities, although essential to consider in the actual selection and the subsequent design process, represent second-order phenomena that will be ignored in analyzing various converter topologies. We will use a generic symbol shown earlier for transistors, regardless of their type, and ignore the need for a specific gate-drive circuitry.

2.6   DESIGN CONSIDERATIONS

In designing any converter, the overall objectives are to optimize its cost, size, weight, energy efficiency, and reliability. With these goals in mind, we will briefly discuss some important considerations as the criteria for selecting the topology and the components in a given application.

2.6.1   Switching Frequency f_s

As discussed in the buck converter example of the previous chapter, a switching power-pole results in a waveform pulsating at a high switching frequency at the current port. The high-frequency components in the pulsating waveform need to be filtered. It is intuitively obvious that the benefits of increasing the switching frequencies lie in reducing the filter component values, L and C, and hence their physical sizes in a converter. (This is true up to a certain value, up to a few hundred kHz, beyond which, for example, magnetic losses in inductors and the internal inductance within capacitors reverse the trend.) Hence, higher switching frequencies allow a higher control bandwidth, as we will see in Chapter 4.

The negative consequences of increasing the switching frequency are in increasing the switching losses in the transistor and the diode, as discussed earlier. Higher switching frequencies also dictate a faster switching of the transistor by appropriately designed gate-drive circuitry, generating greater problems of electromagnetic interference due to higher di/dt and dv/dt that introduce switching noise in the control loop and the rest of the system. We can minimize these problems, at least in DC-DC converters, by adopting a soft-switching topology, discussed in Chapter 10.

2.6.2   Selection of Transistors and Diodes

Earlier, we discussed the voltage and the switching frequency ranges for selecting between MOSFETs and IGBTs. Similarly, the diode types should be chosen appropriately. The voltage rating of these devices is based on the peak voltage $\hat{V}$ in the circuit (including voltage spikes due to parasitic effects during switching). The current ratings should consider the peak current $\hat{I}$ that the devices can handle, which dictates the switching power loss, the RMS current I_{rms} for MOSFETs, which behave as a resistor with $R_{DS(on)}$ in their on-state, and the average current I_{avg} for IGBTs and diodes, which can be approximated to have a constant on-state voltage drop. Safety margins, which depend on the application, dictate that the device ratings be greater than the worst-case stresses by a certain factor.

2.6.3 Magnetic Components

As discussed earlier, the filter inductance value depends on the switching frequency. Inductor design, discussed later in Chapter 9, shows that the physical size of a filter inductor, to a first approximation, depends on a quantity called the area-product (A_p) given by the equation below:

$$A_p = L\,\hat{I}I_{\text{rms}}. \tag{2.6}$$

Increasing L in many topologies reduces the peak and the RMS values (also reducing the transistor and the diode current stresses) but has the negative consequence of increasing the overall inductor size and possibly reducing the control bandwidth.

2.6.4 Capacitor Selection [7]

Capacitors have switching losses designated by equivalent series resistance, ESR, as shown in Figure 2.11. They also have an internal inductance, represented as an equivalent series inductance, ESL. The resonance frequency, the frequency beyond which ESL begins to dominate C, depends on the capacitor type. Electrolytic capacitors offer high capacitance per unit volume but have low resonance frequency. On the other hand, capacitors such as ceramic and metal-polypropylene have relatively high resonance frequency. Therefore, in switch-mode DC power supplies, an electrolytic capacitor with high C may often be paralleled with a ceramic or metal-polyester capacitor to form the output filter capacitor.

2.6.5 Thermal Design [8, 9]

The power dissipated in the semiconductor and the magnetic devices must be removed to limit temperature rise within the device. The reliability of converters and their life expectancy depends on the operating temperatures, which should be well below their maximum allowed values. On the other hand, letting them operate at a high temperature decreases the cost and the size of the heat sinks required. There are several cooling techniques, but for general-purpose applications, converters are often designed for cooling by normal air convection without the use of forced air or liquid.

Semiconductor devices come in a variety of packages, which differ in cost, ruggedness, thermal conduction, and radiation hardness if the application demands it. Figure 2.12a shows a semiconductor device affixed to a heat sink through an isolation pad that is thermally conducting but provides electrical isolation between the device case and the heat sink.

$$\begin{array}{ccc} C & ESL & ESR \end{array}$$

FIGURE 2.11 Capacitor ESR and ESL.

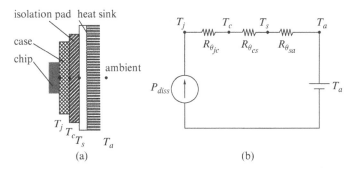

FIGURE 2.12 Thermal design: (a) semiconductor on a heat sink, (b) electrical analog.

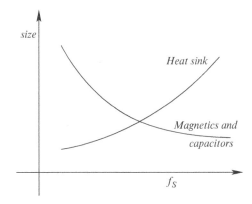

FIGURE 2.13 Size of magnetic components and heat sink as a function of frequency.

An analogy can be drawn with an electrical circuit, as shown in Figure 2.12b, where the power dissipation within the device is represented as a DC current source, thermal resistances offered by various paths are represented as series resistances, and the temperatures at various points as voltages. The data sheets specify various thermal resistance values. The heat transfer mechanism is primarily conduction from the semiconductor device to its case and through the isolation pad. The resulting thermal resistances can be represented by $R_{\theta jc}$ and $R_{\theta cs}$, respectively. From the heat sink to the ambient, the heat transfer mechanisms are primarily convection and radiation, and the thermal resistance can be represented by $R_{\theta sa}$. Based on the electrical analogy, the junction temperature can be calculated as follows for the dissipated power P_{diss}:

$$T_j = T_a + (R_{\theta jc} + R_{\theta cs} + R_{\theta sa})P_{diss}. \tag{2.7}$$

2.6.6 Design Trade-Offs

As a function of switching frequency, Figure 2.13 qualitatively shows the plot of physical sizes for the magnetic components and the capacitors and the heat sink. Based on the present state of the art, optimum values of switching frequencies to minimize the overall size in DC-DC converters using MOSFETs range from 200 kHz to 300 kHz with a slight upward trend.

2.7   THE PWM IC

In the pulse-width modulation of the power pole in a DC-DC converter, a high-speed PWM IC such as the UC3824 from Unitrode/Texas Instruments may be used. Functionally within this PWM IC, the control voltage $v_c(t)$ generated by the controller is compared with a ramp signal v_r of amplitude $\hat{V}_r$, at a constant switching frequency f_s, as shown in Figure 2.14. The output switching signal represents the transistor switching function $q(t)$, which equals 1 if $v_c(t) \geq v_r$ and 0 otherwise. The transistor duty ratio based on Figure 2.14 is given as

$$d(t) = \frac{v_c(t)}{\hat{V}_r}. \tag{2.8}$$

Thus, the control voltage $v_c(t)$ can provide regulation of the average output voltage of the switching power-pole, as discussed further in Chapters 3 and 4.

Many application-specific ICs can be found in [10]. One such IC is UCC28704, which is used to control the transistor in a flyback converter. The IC does not require an optocoupler feedback circuit, instead relying on auxiliary winding to accurately sense the output voltage at the end of demagnetization. The IC also offers quasi-resonant switching where the transistor is turned on at the valley of ringing transistor voltage during discontinuous conduction mode, as explained in Chapter 8. UCC2897A is a controller IC to control a forward converter, also described in Chapter 8. This controller works with an active-clamp forward converter, which eliminates the need for a demagnetization winding to reset the transformer core, making it more compact and efficient.

With the cost of programmable digital controllers rapidly declining, especially ones that support floating-point operation, it is becoming more common to use a programmable digital controller for controlling power electronic converters instead of relying on specific ICs mentioned earlier. One such controller is the Texas Instruments TMS320F280049 DSP. It comes with built-in peripherals such as 14 ADCs, 2 DACs, and 16 PWMs. It has slope generation for internal comparators' DAC for easy implementation of peak current control, as described in Chapter 4. Most PWM peripherals support simple dead-time commutation needed for converters such as the

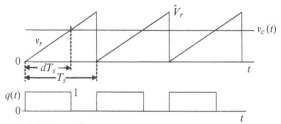

FIGURE 2.14   PWM IC waveforms.

synchronous-rectified buck converter, described in Chapter 3. More complex commutation logic usually requires either external logic ICs or FPGA for programmable logic. The TMS320F280049 DSP has an internal configurable logic block that allows for modifying the PWM signals or input capture signal for converters that require specialized commutation, such as matrix converters.

2.8   HARDWARE PROTOTYPING

All the components mentioned in this chapter are used in the associated Sciamble Power Electronics lab kit, shown in Figure 2.15. The kit consists of three switching power-poles. Two of the switching power-poles are each made up of onsemi's FDMS8090. Each FDMS8090 consists of two 100 V, 10 A Si-MOSFETs. The third switching power-pole is made up of Texas Instruments' LMG5200. LMG5200 is a GaN half-bridge IC, rated for 80 V, 10 A. Both the GaN and the Si switches are driven through a bootstrap driver, discussed in section 2.4.4.

The switches are controlled using TMS320F280049 DSP, mentioned in section 2.7. The DSP is directly programmed using the Sciamble Workbench platform, which allows for developing real-time control using pre-built drag-and-drop toolboxes without any need for C programming or DSP-specific knowledge.

As mentioned in section 2.6.4, the input and output power stages of the lab kit are each decoupled using a single 470 μF capacitor and paralleled with two low ESR 10 μF ceramic capacitors.

Depending on the converter under consideration, the different magnetics sub-circuits, shown in Figure 2.15, are used either with the Si or the GaN power pole, as demonstrated in the following chapters.

The datasheet of all the components used in the Sciamble lab kit, as well as in all the LTspice models in the following chapters, is provided in the Appendix on the accompanying website.

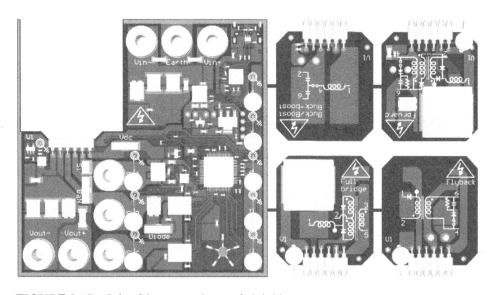

FIGURE 2.15   Sciamble power electronic lab kit.

REFERENCES

1. N. Mohan, T.M. Undeland, and W.P. Robbins, *Power Electronics: Converters, Applications and Design*, 3rd Edition (New York: John Wiley & Sons, 2003).
2. International Rectifier. http://www.irf.com.
3. POWEREX Corporation.: http://www.pwrx.com.
4. Fuji Semiconductor. http://www.fujisemiconductor.com.
5. On Semiconductor. http://www.onsemiconductor.com.
6. Infineon Technologies. http://www.infineon.com.
7. Panasonic Capacitors. http://www.panasonic.com/industrial/electronic-components/capacitive-products.
8. Boyd (Formerly Aavid, Thermal Division of Boyd Corp), for heat sinks, http://boydcorp.com/aavid.html.
9. Bergquist Company for thermal pads.: http://www.bergquistcompany.com.
10. PWM Controller ICs.: https://www.ti.com/power-management/acdc-isolated-dcdc-switching-regulators/pwm-controllers/overview.html.

PROBLEMS

MOSFET in a Power Pole of Figure 2.4a (Problems 2.1 through 2.8)

2.1 A MOSFET is used in a switching power-pole shown in Figure 2.4a. Assume the diode ideal. The operating conditions are as follows: $V_{in} = 42$ V, $I_o = 5$ A, the switching frequency $f_s = 400$ kHz, and the duty ratio $d = 0.3$. Under the operating conditions, the MOSFET has the following switching times: $t_{d(on)} = 50$ ns, $t_{ri} = 20$ ns, $t_{fv} = 15$ ns, $t_{d(off)} = 25$ ns, $t_{rv} = 20$ ns, $t_{fi} = 15$ ns. The on-state resistance of the MOSFET is $R_{DS(on)} = 25$ mΩ. Assume V_{GG} as a step voltage between 0 V and 12 V.
 (a) Draw and label the turn-on and turn-off characteristics of the MOSFET and sketch the MOSFET gate-source voltage v_{GS} waveform.
 (b) Draw and label the diode voltage and current waveforms, assuming it to be ideal.

2.2 Plot the turn-on and the turn-off switching power losses in the MOSFET. Calculate the average switching power loss in the MOSFET.

2.3 Calculate the average conduction loss in the MOSFET.

2.4 Instead of an ideal diode, consider a real diode with a forward voltage drop $V_{FM} = 0.7$ V. Calculate the average forward power loss in the diode.

2.5 In the diode reverse recovery characteristic shown in Figure 2A.1, $t_a = 12$ ns, $t_{rr} = 20$ ns, and $I_{RRM} = 2.5$ A. Calculate the average switching power loss in the diode.

2.6 In the gate-drive circuitry of the MOSFET, assume that the external power supply V_{ext} in Figure 2.9 has a voltage of 12 V. To turn the MOSFET *on* each time, under the conditions given, requires a charge $Q_g = 40$ nC from the 12 V supply. Calculate the average gate-drive power loss, assuming that at turn-off of the MOSFET, no energy is returned to the 12 V supply.

2.7 In this problem, we will calculate new values of the turn-on and the turn-off delay times of the MOSFET, based on the gate driver IC IR2127, as described in Section 2.4.4. This IC is supplied with a voltage $V_{ext} = 12$ V with respect to the MOSFET *source*. Assume this voltage to equal V_{GG}. Also assume the internal series resistance of the MOSFET gate to be zero.

(a) For the turn-on, assume that this driver IC is a voltage source of V_{GG} with an internal resistance of 50 Ω in series. Calculate the turn-on delay time $t_{d(on)}$, assuming that the MOSFET threshold voltage $V_{GS(th)} = 3.5\,\text{V}$. Assume the MOSFET capacitance to be 1800 pF for this calculation.

(b) For turning-off of the MOSFET, assume that this driver IC shorts the MOSFET gate to its source through an internal resistance of 25 Ω. Calculate the turn-off delay time $t_{d(off)}$, assuming that the MOSFET voltage $V_{GS(I_o)} = 4.25$ V. Assume the MOSFET capacitance to be 1800 pF for this calculation.

2.8 The MOSFET losses are the sum of those computed in Problems 2.2 and 2.3. The junction temperature must not exceed 100° C, and the ambient temperature is given as 40° C. From the MOSFET datasheet, $R_{\theta jc} = 1.3\,°\text{C}/\text{W}$. The thermal pad has $R_{\theta cs} = 1.9\,°\text{C}/\text{W}$. Calculate the maximum value of $R_{\theta sa}$ that the heat sink can have.

Simulation Problem

2.9 To achieve the output capacitor in a buck converter, an electrolytic capacitor $C_1 = 680\,\mu\text{F}$ is connected in parallel with a polypropylene capacitor $C_2 = 10\,\mu\text{F}$. Each of these capacitors has the following ESR and ESL as shown in Figure 2.10: $ESR_1 = 0.037\,\Omega$, $ESL_1 = 16\,\text{nH}$, $ESR_2 = 0.04\,\Omega$ and $ESL_1 = 34\,\text{nH}$.

(a) Obtain the frequency response of the admittances associated with C_1 and C_2, and their parallel combination, in terms of their magnitude and the phase angles.

(b) Obtain the resonance frequency for the two capacitors and their combination.

Hint: Look at the phase plot.

APPENDIX 2A   DIODE REVERSE RECOVERY AND POWER LOSSES

In this Appendix, the diode reverse-recovery characteristic is described and the associated power losses are calculated.

2A.1   Average Forward Loss in Diodes

The current flow through a diode results in a forward voltage drop V_{FM} across the diode. The average forward power loss $P_{diode,F}$ in the diode in the circuit of Figure 2.4a can be calculated as

$$P_{diode,F} = (1-d) \cdot V_{FM} I_o, \tag{2A.1}$$

where d is the MOSFET duty ratio in the power pole, and hence the diode conducts for $(1-d)$ portion of each switching time period.

2A.2   Diode Reverse-Recovery Characteristic

Forward current in power diodes, unlike in majority-carrier Schottky diodes, is due to the flow of electrons as well as holes. This current results in an accumulation of electrons in the p-region and of holes in the n-region. The presence of these charges allows a flow of current in the negative direction that sweeps away these excess charges, and the negative current quickly comes to zero, as shown in the plot of Figure 2A.1.

The peak reverse recovery current I_{RRM} and the reverse-recovery charge Q_{rr} shown in Figure 2A.1 depend on the initial forward current I_o and the rate di/dt at which this current decreases.

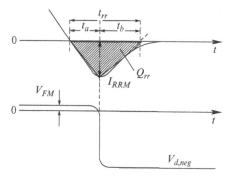

FIGURE 2A.1   Diode reverse-recovery characteristic.

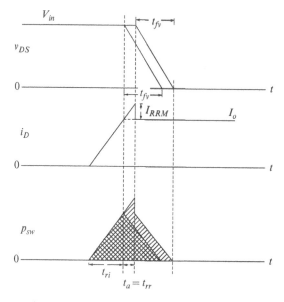

FIGURE 2A.2   Waveforms with diode reverse-recovery current.

2A.3 Diode Switching Losses

The reverse recovery current results in switching losses within the diode when a negative current is flowing beyond the interval t_a in Figure 2A.1, while the diode is blocking a negative voltage $V_{d,neg}$. This switching power loss in the diode can be estimated from the plots of Figure 2A.1 as

$$P_{diode,sw} = \left(\frac{1}{2} I_{RRM} t_b \right) \cdot V_{d,neg} \cdot f_s, \tag{2A.2}$$

where f_s is the switching frequency. In switch-mode power electronics, increase in switching losses due to the diode reverse recovery can be significant, and diodes with ultra-fast reverse recovery characteristics are needed to avoid these from becoming excessive.

2A.4 Diode Switching Losses

The reverse recovery current of the diode also increases the turn-on losses in the associated MOSFET of the power-pole. This is illustrated by an example below.

Example 2.2
In the switching power-pole of Figure 2.4a, $V_{in} = 40$ V and the output current is $I_o = 5$ A. The switching frequency $f_s = 200$ kHz. The MOSFET switching times are $t_{ri} = 15$ ns and $t_{fv} = 15$ ns. The diode snaps off at reverse recovery such that $t_{rr} = t_a = 20$ ns (such that $t_b = 0$) and the peak reverse-recovery current $I_{RRM} = 2$ A. Calculate the additional power loss in the MOSFET due to the diode reverse recovery.

Solution The waveforms are as shown in Figure 2A.2, where the drain current keeps rising for an additional value of $I_{RRM} = 2$ A, beyond $I_o = 5$ A, during $t_a = t_{rr} = 20$ ns. Beyond this interval, the drain-source voltage falls during the interval $t_{fv} = 15$ ns.

With a near-ideal diode, as soon as the drain current reaches I_o, the MOSFET voltage v_{DS} begins to drop linearly in an interval t_{fv}, and the corresponding switching loss in the MOSFET is shown cross-hatched:

$$P_{sw,ideal} = \frac{1}{2} \times V_{in} \times I_o \times (t_{ri} + t_{fv}) \times f_s = 0.6 \, \text{W}. \tag{2A.3}$$

With the diode with a reverse-recovery current, the switching loss in the MOSFET is

$$P_{sw,RR} = \frac{1}{2} \times V_{in} \times (I_o + I_{RRM}) \times (t_{ri} + t_{rr}) \times f_s + \frac{1}{2} \times V_{in} \times I_o \times t_{fv} \times f_s = 1.28 \, \text{W}. \tag{2A.4}$$

Therefore, the additional power loss in the MOSFET is 0.68 W, which more than doubles the power loss in it.

3

SWITCH-MODE DC-DC CONVERTERS: SWITCHING ANALYSIS, TOPOLOGY SELECTION, AND DESIGN

In Chapter 1, we discussed various applications of power electronics, including those for energy sustainability. Some of these applications, such as harnessing solar energy using photovoltaics, use of fuel cells, and energy storage in batteries, require DC-DC converters to convert voltages and currents from one DC level to another and to operate these systems optimally. In addition to these direct applications, DC-DC converters form the basic block of conversion between AC and DC voltages, required in applications such as harnessing wind energy and efficiently adjusting the speed of drives in transportation and compressor systems for increasing energy efficiency.

3.1 DC-DC CONVERTERS [1]

Figure 3.1a shows buck, boost, and buck-boost DC-DC converters by a block diagram and how the pulse-width modulation process regulates their output voltage. The design of the feedback controller is the subject of the next chapter. The power flow through these converters is in only one direction. Thus, their voltages and currents remain unipolar and unidirectional, as shown in Figure 3.1b. Based on these converters, several transformer-isolated DC-DC converter topologies, which are used in all types of electronics equipment, are discussed in Chapter 8.

3.2 SWITCHING POWER-POLE IN DC STEADY STATE

All the converters that we will discuss consist of a switching power-pole that was introduced in Chapter 1 and is redrawn in Figure 3.2a. In these converter circuits in DC steady state, the input voltage and the output load are assumed constant. The switching

Power Electronics A First Course: Simulations and Laboratory Implementations, Second Edition.
Ned Mohan and Siddharth Raju.
© 2023 John Wiley & Sons, Inc. Published 2023 by John Wiley & Sons, Inc.
Companion Website: www.wiley.com/go/mohan/powerelectronics2e

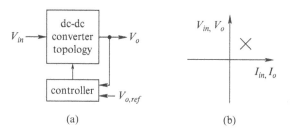

(a)   (b)

FIGURE 3.1   Regulated switch-mode DC power supplies.

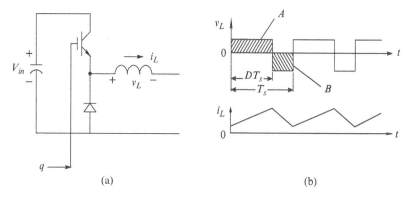

(a)   (b)

FIGURE 3.2   Switching power-pole as the building block of DC-DC converters.

power-pole operates with a transistor switching function $q(t)$, whose waveform repeats, unchanged from one cycle to the next, and the corresponding switch duty ratio remains constant at its steady-state DC value D. Therefore, all waveforms associated with the power pole repeat with the switching time period T_s in the DC steady state, where the basic principles described below are extremely useful for analysis purposes.

First, let us consider the voltage and current of the inductor associated with the power pole. The inductor current depends on the pulsating voltage waveform, as shown in Figure 3.2b. The inductor voltage and current are related by the conventional differential equation, which can be expressed in the integral form as follows:

$$v_L = L\frac{di_L}{dt} \Rightarrow i_L(t) = \frac{1}{L}\int_\tau v_L \cdot d\tau, \tag{3.1}$$

where τ is a variable of integration representing time. For simplicity, we will consider the first time period starting with $t = 0$ in Figure 3.2b. Using the integral form in Equation (3.1), the inductor current at a time t can be expressed in terms of its initial value $i_L(0)$ as:

$$i_L(t) = i_L(0) + \frac{1}{L}\int_0^t v_L \cdot d\tau. \tag{3.2}$$

In the DC steady state, the waveforms of all circuit variables must repeat with the switching frequency time period T_s, resulting in the following conclusions from Equation (3.2):

1. The inductor current waveforms repeat with T_s, and therefore in Equation (3.2)

$$i_L(T_s) = i_L(0). \tag{3.3}$$

2. Integrating over one switching time period T_s in Equation (3.2) and using Equation (3.3) show that the inductor voltage integral over T_s is zero. This leads to the conclusion that the average inductor voltage, averaged over T_s, is zero:

$$\frac{1}{L}\int_0^{T_s} v_L \cdot d\tau = 0 \quad \Rightarrow V_L = \frac{1}{T}\left(\underbrace{\int_0^{DT_s} v_L \cdot d\tau}_{area\ A} + \underbrace{\int_{DT_s}^{T_s} v_L \cdot d\tau}_{area\ B}\right) = 0. \tag{3.4}$$

In Figure 3.2a, the area A in volt-seconds, which causes the current to rise, equals in magnitude the negative area B, which causes the current to decline to its initial value.

Example 3.1
If the current waveform in steady state in an inductor of $50\,\mu H$ is as shown in Figure 3.3a, calculate the inductor voltage waveform $v_L(t)$.

Solution During the current rise time, $\dfrac{di}{dt} = \dfrac{(4-3)}{3\mu} = \left(\dfrac{1}{3\mu}\right)\dfrac{A}{s}$. Therefore,

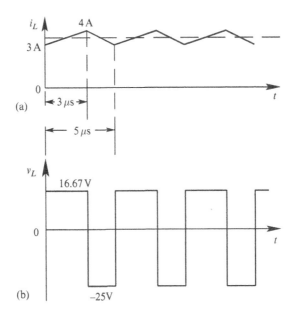

(a)

(b)

FIGURE 3.3 Example 3.1.

$$v_L = L\frac{di}{dt} = 50\mu \times \frac{1}{3\mu} = 16.67\,\text{V}.$$

During the current fall time, $\dfrac{di}{dt} = \dfrac{(3-4)}{2\mu} = \left(-\dfrac{1}{2\mu}\right)\dfrac{A}{s}$. Hence,

$$v_L = L\frac{di}{dt} = 50\mu \times \left(-\frac{1}{2\mu}\right) = -25\,\text{V}.$$

Therefore, the inductor voltage waveform is as shown in Figure 3.3b.

The above analysis applies to any inductor in a switch-mode converter circuit operating in a DC steady state. By analogy, a similar analysis applies to any capacitor in a switch-mode converter circuit operating in the DC steady state as follows: The capacitor voltage and current are related by the conventional differential equation, which can be expressed in the integral form as follows:

$$i_C = C\frac{dv_C}{dt} \quad \Rightarrow \quad v_C(t) = \frac{1}{C}\int_\tau i_C \cdot d\tau, \tag{3.5}$$

where τ is a variable of integration representing time. Using the integral form of Equation (3.5), the capacitor voltage at a time t can be expressed in terms of its initial value $v_C(0)$ as:

$$v_C(t) = v_C(0) + \frac{1}{C}\int_0^t i_C \cdot d\tau. \tag{3.6}$$

In the DC steady state, the waveforms of all circuit variables must repeat with the switching frequency time period T_s, resulting in the following conclusions from Equation (3.6):

1. The capacitor voltage waveform repeats with T_s, and therefore in Equation (3.6)

$$v_C(T_s) = v_C(0). \tag{3.7}$$

2. Integrating over one switching time period T_s in Equation (3.6) and using Equation (3.7) show that the capacitor current integral over T_s is zero, which leads to the conclusion that the average capacitor current, averaged over T_s, is zero:

$$\frac{1}{C}\int_0^{T_s} i_C \cdot d\tau = 0 \quad \Rightarrow \quad I_C = \frac{1}{T_s}\int_0^{T_s} i_C \cdot d\tau = 0 \tag{3.8}$$

Example 3.2

The capacitor current i_C, shown in Figure 3.4a, is flowing through a $100\,\mu\text{F}$ capacitor. Calculate the peak-peak ripple in the capacitor voltage waveform due to this ripple current.

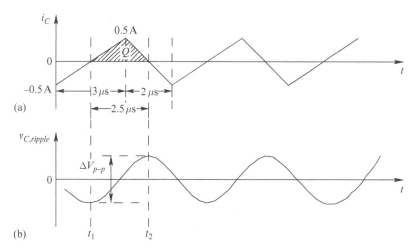

FIGURE 3.4 Example 3.2.

Solution For the given capacitor current waveform, the capacitor voltage waveform, as shown in Figure 3.4b, is at its minimum at time t_1, before which the capacitor current has been negative. This voltage waveform reaches its peak at time t_2, beyond which the current becomes negative.

The hatched area in Figure 3.4a equals the charge $Q = \int_{t_1}^{t_2} i_C \,.dt = \frac{1}{2} \times 0.5 \times 2.5\,\mu C = 0.625\,\mu C$.

Using Equation (3.6), the peak-peak ripple in the capacitor voltage is

$$\Delta V_{p-p} = \frac{Q}{C} = 6.25\,mV.$$

In addition to the above two conclusions, it is important to recognize that in DC steady state, just as with instantaneous quantities, Kirchhoff's voltage and current laws apply to average quantities as well. In the DC steady state, average voltages sum to zero in a circuit loop, and average currents sum to zero at a node:

$$\sum_k V_k = 0, \tag{3.9}$$

$$\sum_k I_k = 0. \tag{3.10}$$

3.3   SIMPLIFYING ASSUMPTIONS

To gain a clear understanding of the DC steady state, we will first make certain simplifying assumptions by ignoring the second-order effects listed below, and later on, we will include them for accuracy:

1. Transistors, diodes, and other passive components are all ideal unless explicitly stated. For example, we will ignore the inductor equivalent series resistance.
2. The input is a pure DC voltage V_{in}.

3. Design specifications require the ripple in the output voltage to be very small. Therefore, we will initially assume that the output voltage is purely DC without any ripple, that is $v_o(t) \simeq V_o$, and later calculate the ripple in it.
4. The current at the current port of the power pole through the series inductor flows continuously, resulting in a continuous conduction mode, CCM (the discontinuous conduction mode, DCM, is analyzed later on).

It is, of course, possible to analyze a switching circuit in detail without making the above simplifying assumptions, as we will do in LTspice-based computer simulations. However, the two-step approach followed here, where the analysis is first carried out by neglecting the second-order effects and adding them later on, provides a deeper insight into converter operation and the design trade-offs.

3.4 COMMON OPERATING PRINCIPLES

In all three converters that we will analyze, the inductor associated with the switching power-pole acts as an energy transfer means from the input to the output. Turning on the transistor of the power pole increases the inductor energy by a certain amount, drawn from the input source, which is transferred to the output stage during the off interval of the transistor. In addition to this inductive energy-transfer means, depending on the converter, there may be additional energy transfer directly from the input to the output, as discussed in the following sections.

3.5 BUCK CONVERTER SWITCHING ANALYSIS IN DC STEADY STATE

A buck converter is shown in Figure 3.5a, with the transistor and the diode making up the bi-positional switch of the power pole. The equivalent series resistance (ESR) of the capacitor will be ignored. Turning on the transistor increases the inductor current in the sub-circuit of Figure 3.5b. When the transistor is turned off, the inductor current freewheels through the diode, as shown in Figure 3.5c.

For a given transistor switching function waveform $q(t)$ shown in Figure 3.5d with a switch duty ratio D in steady state, the waveform of the voltage v_A at the current port follows $q(t)$ as shown. In Figure 3.5d, integrating v_A over T_s, the average voltage V_A equals DV_{in}. Recognizing that the average inductor voltage is zero (Equations 3.4) and the average voltages in the output loop sum to zero (Equations 3.9),

$$V_o = V_A = DV_{in}. \tag{3.11}$$

The inductor voltage v_L pulsates between two values, $(V_{in} - V_o)$ and $(-V_o)$, as plotted in Figure 3.5d. Since the average inductor voltage is zero, the volt-second areas during two subintervals are equal in magnitude and opposite in sign. In DC steady state, the inductor current can be expressed as the sum of its average and the ripple component:

$$i_L(t) = I_L + i_{L,ripple}(t), \tag{3.12}$$

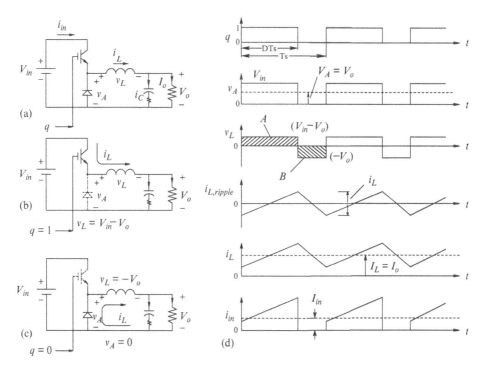

FIGURE 3.5   Buck DC-DC converter.

where the average current depends on the output load, and the ripple component is dictated by the waveform of the inductor voltage v_L in Figure 3.5d. As shown in Figure 3.5d, the ripple component consists of linear segments, rising when v_L is positive and falling when v_L is negative. The peak-peak ripple can be calculated as follows, using either area A or B:

$$\Delta i_L = \frac{1}{L}\underbrace{(V_{in} - V_o)DT_s}_{Area\ A} = \frac{1}{L}\underbrace{V_o(1- D)T_s}_{Area\ B}. \tag{3.13}$$

This ripple component is plotted in Figure 3.5d. Since the average capacitor current is zero in DC steady state, the average inductor current equals the output load current by Kirchhoff's current law applied at the output node:

$$I_L = I_o = \frac{V_o}{R}. \tag{3.14}$$

The inductor current waveform is shown in Figure 3.5d by superposing the average and the ripple components.

Next, we will calculate the ripple current through the output capacitor. In practice, the filter capacitor is large enough to achieve the output voltage, nearly DC $(v_o(t) \approx V_o)$. Therefore, to the ripple-frequency current, the path through the capacitor offers much smaller impedance than through the load resistance, hence justifying the assumption that the ripple component of the inductor current flows entirely through the capacitor. That is, in Figure 3.5a,

$$i_C(t) \approx i_{L,ripple}(t) \tag{3.15}$$

In practice, in a capacitor, the voltage drops across its equivalent series resistance (ESR) and the equivalent series inductance (ESL) dominate over the voltage drop $\frac{1}{C}\int i_C \, dt$ across C, given by Equation (3.6). The capacitor current i_C, equal to $i_{L,ripple}$ in Figure 3.5d, can be used to calculate the ripple in the output voltage.

The input current i_{in} pulsates, equal to i_L when the transistor is on, and otherwise zero, as plotted in Figure 3.5d. An input L-C filter is often needed to prevent the pulsating current from being drawn from the input DC source. The average value of the input current in Figure 3.5d is

$$I_{in} = D I_L = D I_o \quad \text{(using Eq.3.14).} \tag{3.16}$$

Using Equations (3.11) and (3.16), we can confirm that the input power equals the output power, as it should, in this idealized converter:

$$V_{in} I_{in} = V_o I_o \tag{3.17}$$

Equation (3.11) shows that the voltage conversion ratio of buck converters in the continuous conduction mode (CCM) depends on D but is independent of the output load. If the output load decreases (that is, if the load resistance increases) to the extent that the inductor current becomes discontinuous, then the input-output relationship in CCM is no longer valid, and, if the duty ratio D were to be held constant, the output voltage in the discontinuous conduction mode would rise above that given by Equation (3.11). The discontinuous conduction mode will be considered fully in section 3.15.

Example 3.3
In the buck DC-DC converter shown in Figure 3.5a, $L = 24\,\mu\text{H}$. It is operating in DC steady state under the following conditions: $V_{in} = 20\,\text{V}$, $D = 0.6$, $P_o = 14\,\text{W}$, and $f_s = 200\,\text{kHz}$. Assuming ideal components, calculate and draw the waveforms shown earlier in Figure 3.5d.

Solution   With $f_s = 200\,\text{kHz}$, $T_s = 5\,\mu\text{s}$ and $T_{on} = DT_s = 3\,\mu\text{s}$, $V_o = DV_{in} = 12\,\text{V}$.

The inductor voltage v_L fluctuates between $(V_{in} - V_o) = 8\,\text{V}$ and $(-V_o) = -12\,\text{V}$, as shown in Figure 3.6.

Therefore, from Equation (3.13), the ripple in the inductor current is $\Delta i_L = 1\,\text{A}$. The average inductor current is $I_L = I_o = P_o/V_o = 1.167\,\text{A}$. Therefore, $i_L = I_L + i_{L,ripple}$, as shown in Figure 3.6. When the transistor is on, $i_{in} = i_L$, and otherwise it zero. The average input currents is $I_{in} = D I_o = 0.7\,\text{A}$.

3.5.1   Simulation and Hardware Prototyping

The simulation of a non-ideal buck converter is demonstrated by means of an example:

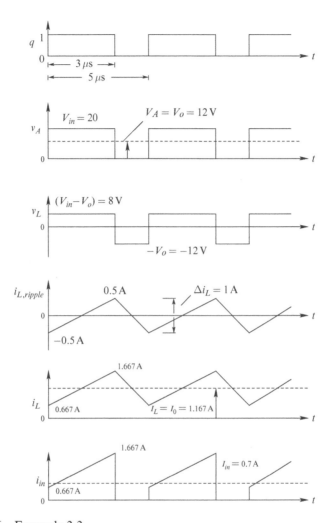

FIGURE 3.6   Example 3.3.

Example 3.4

In the buck DC-DC converter shown in Figure 3.5a, $L = 68\,\mu\text{H}$, $C = 490\,\mu\text{F}$, and $R = 8\,\Omega$. It is operating in DC steady state under the following conditions: $V_{in} = 15\,\text{V}$, $D = 0.7$, and $f_s = 100\,\text{kHz}$. For the switch and the diode, use the parameters given in the Appendix of Chapter 2. Simulate this converter using LTspice.

Solution   The simulation file used in this example is available on the accompanying website. The LTspice model is shown in Figure 3.7, and the steady-state waveforms from the simulation of this converter are shown in Figure 3.8.

The Workbench model for implementing the above example in hardware using the Sciamble lab kit is as shown in Figure 3.9.

The steady-state waveforms from running the buck converter using the Sciamble laboratory kit are shown in Figure 3.10. The step-by-step procedure for re-creating the above hardware implementation is presented in [2].

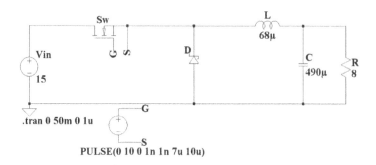

FIGURE 3.7 LTspice model.

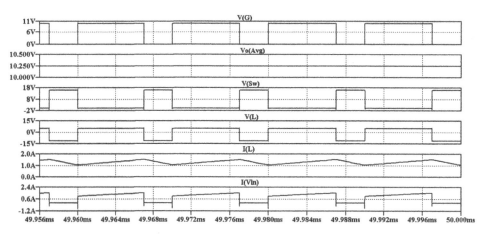

FIGURE 3.8 LTspice simulation results.

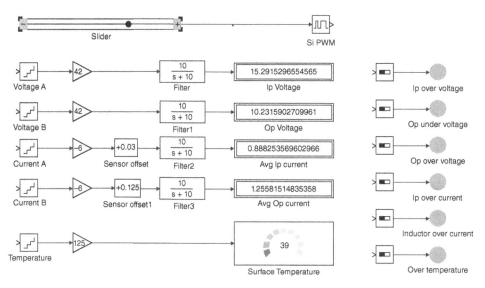

FIGURE 3.9 Workbench model.

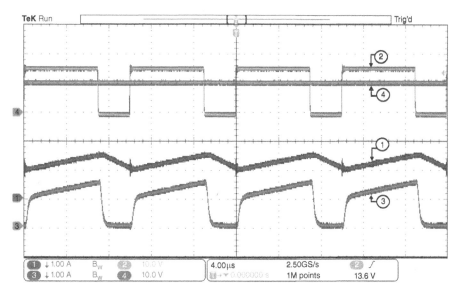

FIGURE 3.10 Workbench hardware results: (1) inductor current, (2) switch-node voltage, (3) input current, and (4) output voltage.

3.6 BOOST CONVERTER SWITCHING ANALYSIS IN DC STEADY STATE

A Boost converter is shown in Figure 3.11a. It is conventional to show power flow from left to right. To follow this convention, the circuit of Figure 3.11a is flipped and drawn in Figure 3.11b. The output stage consists of the output load and a large filter capacitor that is used to minimize the output voltage ripple. This capacitor at the output initially gets charged to a voltage equal to V_{in} through the diode.

Compared to buck converters, the boost converters have two major differences:

1. Power flow is from a lower voltage DC input to the higher load voltage in the opposite direction through the switching power-pole. Hence, the current direction through the series inductor of the power pole is chosen as shown, opposite to that in a buck converter, and this current remains positive in the continuous conduction mode.
2. In the switching power-pole, the bi-positional switch is realized using a transistor and a diode that are placed as shown in Figure 3.7a. Across the output, a filter capacitor C is placed, which forms the voltage port and minimizes the output ripple voltage.

In a boost converter, turning on the transistor in the bottom position applies the input voltage across the inductor such that v_L equals V_{in}, as shown in Figure 3.12a, and i_L linearly ramps up, increasing the energy in the inductor. Turning off the transistor forces the inductor current to flow through the diode, as shown in Figure 3.12b, and some of the inductively stored energy is transferred to the output stage that consists of the filter capacitor and the output load across it.

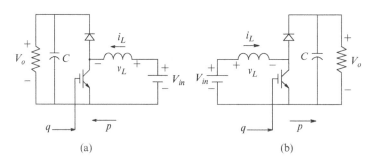

FIGURE 3.11   Boost DC-DC converter.

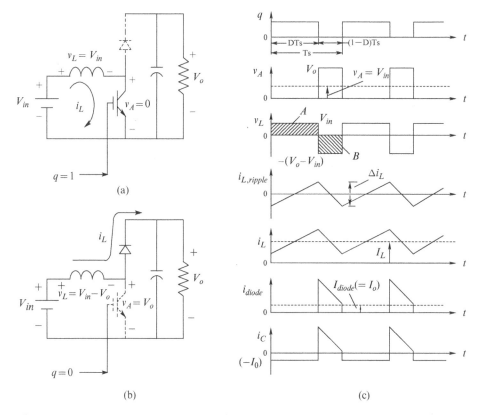

FIGURE 3.12   Boost converter: operation and waveforms.

The transistor switching function is shown in Figure 3.12c, with a steady-state duty ratio D. Because of the transistor in the bottom position in the power pole, the resulting v_A waveform is as plotted in Figure 3.12c. Since the average voltage across the inductor in the DC steady state is zero, the average voltage V_A equals the input voltage V_{in}. The inductor voltage v_L pulsates between two values: V_{in} and $-(V_o - V_{in})$ as plotted in Figure 3.12c. Since the average inductor voltage is zero, the volt-second areas during the two subintervals are equal in magnitude and opposite in sign.

The input/output voltage ratio can be obtained either by the waveform of v_A or v_L in Figure 3.12c. Using the inductor voltage waveform whose average is zero in DC steady state,

$$V_{in}(DT_s) = (V_o - V_{in})(1 - D)T_s. \tag{3.18}$$

Hence,

$$\frac{V_o}{V_{in}} = \frac{1}{1 - D} \qquad (V_o > V_{in}). \tag{3.19}$$

The inductor current waveform consists of its average value, which depends on the output load, and a ripple component, which depends on v_L:

$$i_L(t) = I_L + i_{L,ripple}(t), \tag{3.20}$$

where as shown in Figure 3.12c, $i_{L,ripple}\left(= \frac{1}{L}\int v_L \cdot d\tau\right)$, whose average value is zero, consists of linear segments, rising when v_L is positive and falling when v_L is negative. The peak-peak ripple can be calculated by using either area A or B:

$$\Delta i_L = \frac{1}{L}\underbrace{V_{in}(DT_s)}_{Area\ A} = \frac{1}{L}\underbrace{(V_o - V_{in})(1 - D)T_s}_{Area\ B}. \tag{3.21}$$

In a boost converter, the inductor current equals the input current, whose average can be calculated from the output load current by equating the input and the output powers:

$$V_{in}I_{in} = V_o I_o. \tag{3.22}$$

Hence, using Equation (3.19) and $I_o = V_o / R$,

$$I_L = I_{in} = \frac{V_o}{V_{in}}I_o = \frac{I_o}{1 - D} = \frac{1}{1 - D}\frac{V_o}{R}. \tag{3.23}$$

The inductor current waveform is shown in Figure 3.12c, superposing its average and the ripple components.

The current through the diode equals 0 when the transistor is on; otherwise, it equals i_L, as plotted in Figure 3.12c. In the DC steady state, the average capacitor current I_C is zero, and therefore the average diode current equals the output current I_o. In practice, the filter capacitor is large to achieve the output voltage of nearly DC $(v_o(t) \approx V_o)$. Therefore, to the ripple-frequency component in the diode current, the path through the capacitor offers a much smaller impedance than through the load resistance, hence justifying the assumption that the ripple component of the diode current flows entirely through the capacitor. That is,

$$i_C(t) \simeq i_{diode,ripple}(t) = i_{diode} - I_o. \tag{3.24}$$

In practice, the voltage drops across the capacitor ESR and the ESL dominate over the voltage drop $\frac{1}{C} \int i_C \, dt$ across C. The plot of i_C in Figure 3.8c can be used to calculate the ripple in the output voltage.

Example 3.5

In a boost converter, shown in Figure 3.11b, the inductor current has $\Delta i_L = 2\,\text{A}$. It is operating in DC steady state under the following conditions: $V_{in} = 5\,\text{V}$, $V_o = 12\,\text{V}$, $P_o = 11\,\text{W}$, and $f_s = 200\,\text{kHz}$. (a) Assuming ideal components, calculate L and draw the waveforms as shown in Figure 3.12c.

Solution   From Equation (3.19), the duty ratio $D = 0.583$. With $f_s = 200\,\text{kHz}$, $T_s = 5\,\mu s$ and $T_{on} = DT_s = 2.917\,\mu s$. v_L fluctuates between $V_{in} = 5\,\text{V}$ and $-(V_o - V_{in}) = -7\,\text{V}$. Using the conditions during the transistor on-time, from Equation (3.21),

$$L = \frac{V_{in}}{\Delta i_L} DT_s = 7.29\,\mu\text{H}.$$

The average inductor current is $I_L = I_{in} = P_{in}(= P_o)/V_{in} = 2.2\,\text{A}$, and $i_L = I_L + i_{L,ripple}$, as shown in Figure 3.13. When the transistor is on, the diode current is zero; otherwise $i_{diode} = i_L$. The average diode current is equal to the average output current:

$$I_{diode} = I_o = (1-D)I_{in} = 0.917\,\text{A}.$$

The capacitor current is $i_C = i_{diode} - I_o$. When the transistor is on, the diode current is zero and $i_C = -I_o = -0.917\,\text{A}$. The capacitor current jumps to a value of $2.283\,\text{A}$ and drops to $-0.917\,\text{A}$.

The above analysis shows that the voltage conversion ratio (Equation 3.19) of boost converters in CCM depends on $1/(1-D)$, and is independent of the output load, as shown in Figure 3.14. If the output load decreases to the extent that the average inductor current becomes less than the critical value $I_{L,crit}$, the inductor current becomes discontinuous, and in this discontinuous conduction mode (DCM), the input-output relationship of CCM is no longer valid, as shown in Figure 3.14.

If the duty ratio D were to be held constant, as shown in Figure 3.14, the output voltage could rise to dangerously high levels in DCM; this case is fully considered in Section 3.15.

3.6.1   Simulation and Hardware Prototyping

The simulation of a non-ideal boost converter is demonstrated by means of an example:

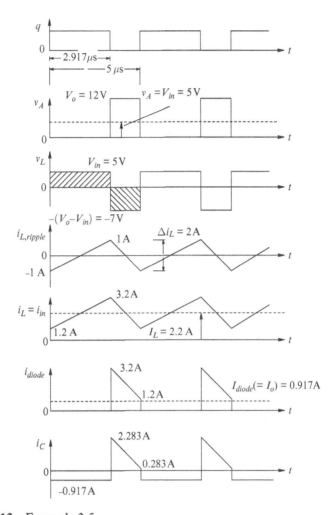

FIGURE 3.13 Example 3.5.

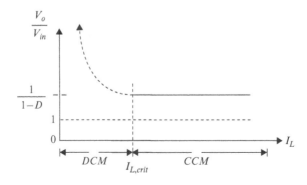

FIGURE 3.14 Boost converter: voltage transfer ratio.

Example 3.6

In the boost DC-DC converter shown in Figure 3.11b, $L = 68\,\mu H$, $C = 490\,\mu F$, and $R = 39\,\Omega$. It is operating in DC steady state under the following conditions: $V_{in} = 15\,V$, $D = 0.3$, and $f_s = 100\,kHz$. For the switch and the diode, use the parameters given in the Appendix of Chapter 2. Simulate this converter using LTspice.

Solution   The simulation file used in this example is available on the accompanying website. The LTspice model is shown in Figure 3.15, and the steady-state waveforms from the simulation of this converter are shown in Figure 3.16.

The Workbench model for implementing the above example in hardware using the Sciamble lab kit is shown in Figure 3.17.

The steady-state waveforms from running the boost converter using the Sciamble laboratory kit are shown in Figure 3.18. The step-by-step procedure for re-creating the above hardware implementation is presented in [3].

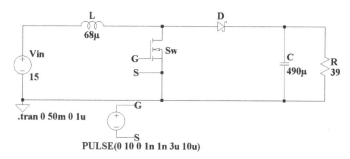

FIGURE 3.15   LTspice model.

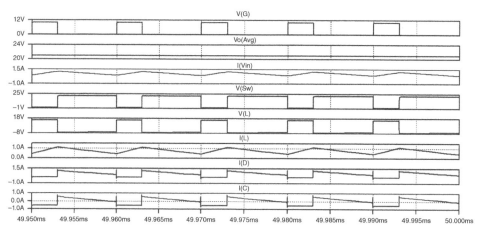

FIGURE 3.16   LTspice simulation results.

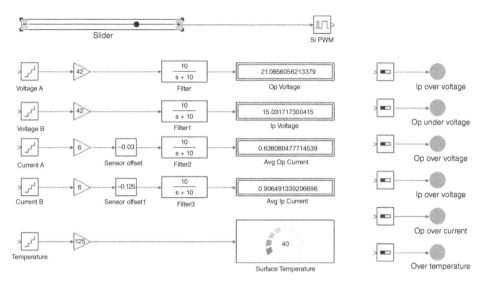

FIGURE 3.17   Workbench model.

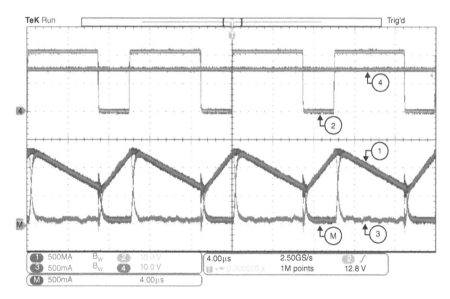

FIGURE 3.18   Workbench hardware results: (1) input current, (2) switch-node voltage, (3) diode current, (4) output voltage, and (M) switch current.

3.7   BUCK-BOOST CONVERTER ANALYSIS IN DC STEADY STATE

Buck-boost converters allow the output voltage to be greater or lower than the input voltage based on the switch duty ratio D. A buck-boost converter is shown in Figure 3.19a, where the switching power-pole is implemented as shown. Conventionally, to make the power flow from left to right, buck-boost converters are drawn as in Figure 3.19b.

As shown in Figure 3.20a, turning on the transistor applies the input voltage across the inductor such that v_L equals V_{in}, and the current linearly ramps up, increasing the energy in the inductor. Turning off the transistor results in the inductor current flowing through the diode, as shown in Figure 3.20b, transferring to the output the incremental energy in the inductor which was accumulated during the previous transistor state.

The transistor-switching function is shown in Figure 3.20c, with a steady-state duty ratio D. The resulting v_A waveform is as plotted. Since the average voltage across the inductor in the DC steady state is zero, the average V_A equals the output voltage V_o.

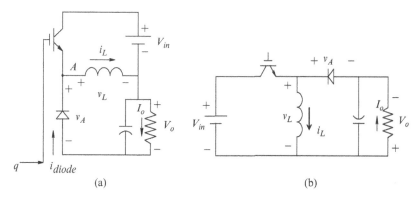

FIGURE 3.19   Buck-boost DC-DC converter.

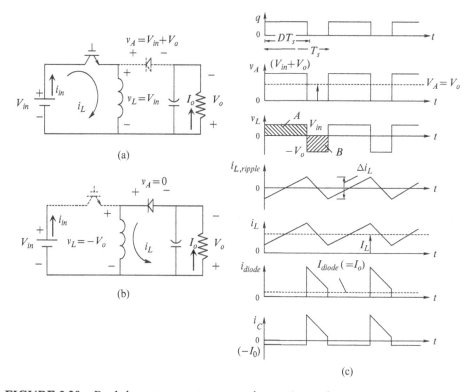

FIGURE 3.20   Buck-boost converter: operation and waveforms.

The inductor voltage pulsates between two values: V_{in} and $-V_o$, as plotted in Figure 3.20c. Since the average inductor voltage is zero, the volt-second areas during the two subintervals are equal in magnitude and opposite in sign.

The input/output voltage ratio can be obtained either by the waveform of v_A or v_L in Figure 3.20c. Using the v_L waveform, whose average is zero in the DC steady state,

$$DV_{in} = (1 - D)V_o. \tag{3.25}$$

Hence,

$$\frac{V_o}{V_{in}} = \frac{D}{1 - D}. \tag{3.26}$$

The inductor current consists of an average value, which depends on the output load, and a ripple component, which depends on v_L:

$$i_L(t) = I_L + i_{L,ripple}(t), \tag{3.27}$$

where as shown in Figure 3.20c, $i_{L,ripple}\left(= \frac{1}{L}\int v_L \cdot d\tau\right)$, whose average value is zero, consists of linear segments, rising when v_L is positive and falling when v_L is negative. The peak-peak ripple can be calculated by using either area A or B,

$$\Delta i_L = \frac{1}{L}\underbrace{V_{in}(DT_s)}_{Area\ A} = \frac{1}{L}\underbrace{V_o(1-D)T_s}_{Area\ B} \tag{3.28}$$

Applying Kirchhoff's current law in Figure 3.19a or 3.19b, the average inductor current equals the sum of the average input current and the average output current (note that the average capacitor current is zero in DC steady state),

$$I_L = I_{in} + I_o. \tag{3.29}$$

Equating the input and the output powers,

$$V_{in}I_{in} = V_oI_o, \tag{3.30}$$

and using Equation (3.26),

$$I_{in} = \frac{V_o}{V_{in}}I_o = \frac{D}{1-D}I_o. \tag{3.31}$$

Hence, using Equations (3.29) and (3.31),

$$I_L = I_{in} + I_o = \frac{1}{1-D}I_o = \frac{1}{1-D}\frac{V_o}{R}. \tag{3.32}$$

Superposing the average and the ripple components, the inductor current waveform is shown in Figure 3.20c.

The diode current is zero, except when it conducts the inductor current, as plotted in Figure 3.20c. In the DC steady state, the average current I_C through the capacitor is zero, and therefore by Kirchhoff's current law, the average diode current equals the output current. In practice, the filter capacitor is large to achieve the output voltage nearly DC ($v_o(t) \simeq V_o$). Therefore, to the ripple-frequency current, the path through the capacitor offers a much smaller impedance than through the load resistance, hence justifying the assumption that the ripple component of the diode current flows entirely through the capacitor. That is,

$$i_C(t) \simeq i_{diode,ripple}(t). \tag{3.33}$$

In practice, the voltage drops across the capacitor ESR and the ESL dominate over the voltage drop $\dfrac{1}{C}\int i_C\,dt$ across C. The plot of i_C in Figure 3.20c can be used to calculate the ripple in the output voltage.

Example 3.7
The buck-boost converter of Figure 3.19b is operating in DC steady state under the following conditions: $V_{in} = 14\,\text{V}$, $V_o = 42\,\text{V}$, $P_o = 21\,\text{W}$, $\Delta i_L = 1.8\,\text{A}$, and $f_s = 200\,\text{kHz}$. Assuming ideal components, calculate L and draw the waveforms as shown in Figure 3.20c.

Solution   From Equation (3.26), $D = 0.75$. $T_s = 1/f_s = 5\,\mu s$ and $T_{on} = DT_s = 3.75\,\mu s$, as shown in Figure 3.13. The inductor voltage v_L fluctuates between $V_{in} = 14\,\text{V}$ and $-V_o = -42\,\text{V}$. Using Equation (3.28),

$$L = \frac{V_{in}}{\Delta i_L}DT_s = 29.17\,\mu\text{H}.$$

The average input current is $I_{in} = P_{in}(= P_o)/V_{in} = 1.5\,\text{A}$, and $I_o = P_o/V_o = 0.5\,\text{A}$. Therefore, $I_L = I_{in} + I_o = 2\,\text{A}$. When the transistor is on, the diode current is zero; otherwise, $i_{diode} = i_L$. The average diode current is equal to the average output current: $I_{diode} = I_o = 0.5\,\text{A}$. The capacitor current is $i_C = i_{diode} - I_o$. When the transistor is on, the diode current is zero and $i_C = -I_o = -0.5\,\text{A}$. The capacitor current jumps to a value of 2.4 A and drops to $1.1 - 0.5 = 0.6\,\text{A}$, as shown in Figure 3.21.

The above analysis shows that the voltage conversion ratio (Equation 3.26) of buck-boost converters in CCM depends on $D/(1-D)$ and is independent of the output load, as shown in Figure 3.22. If the output load decreases to the extent that the average inductor current becomes less than the critical value $I_{L,crit}$, the inductor current becomes discontinuous, and in this discontinuous conduction mode (DCM), the input-output relationship of CCM is no longer valid, as shown in Figure 3.22. If the duty ratio D were to be held constant as shown in Figure 3.22, the output voltage could rise to dangerously high levels in DCM; this case is fully considered in Section 3.23.

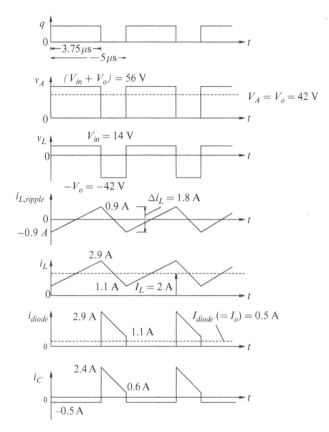

FIGURE 3.21   Example 3.7.

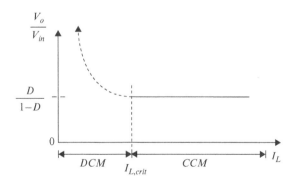

FIGURE 3.22   Buck-boost converter: voltage transfer ratio.

3.7.1 Simulation and Hardware Prototyping

The simulation of a non-ideal buck-boost converter is demonstrated by means of an example:

Example 3.8

In the buck-boost DC-DC converter shown in Figure 3.19b, $L = 68\,\mu H$, $C = 490\,\mu F$, and $R = 25\,\Omega$. It is operating in DC steady state under the following conditions: $V_{in} = 15\,V$, $D = 0.3$, and $f_s = 100\,kHz$. For the switch and the diode, use the parameters given in the Appendix of Chapter 2. Simulate this converter using LTspice.

Solution   The simulation file used in this example is available on the accompanying website. The LTspice model is shown in Figure 3.23, and the steady-state waveforms from the simulation of this converter are shown in Figure 3.24.

The Workbench model for implementing the above example in hardware using the Sciamble lab kit is as shown in Figure 3.25.

The steady-state waveforms from running the buck-boost converter using the Sciamble laboratory kit are shown in Figure 3-26. The step-by-step procedure for re-creating the above hardware implementation is presented in [4].

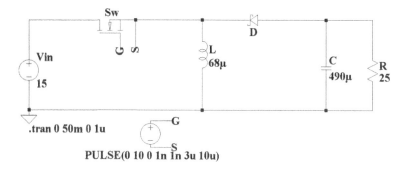

FIGURE 3.23   LTspice model.

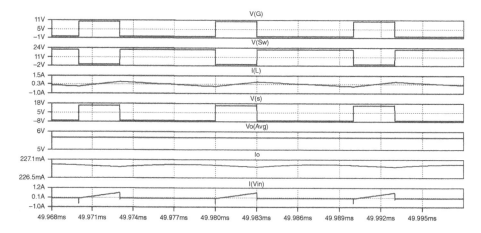

FIGURE 3.24   LTspice simulation results.

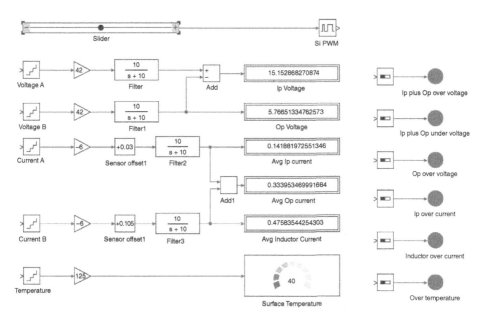

FIGURE 3.25   Workbench model.

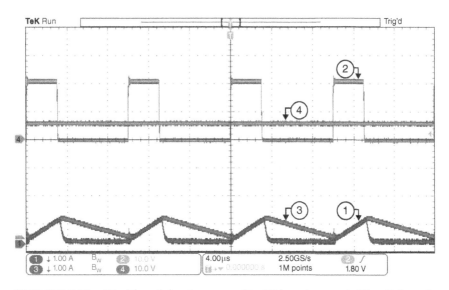

FIGURE 3.26   Workbench hardware results: (1) input current, (2) switch-node voltage, (3) inductor current, and (4) output voltage.

3.7.2   Other Buck-Boost Topologies

There are two variations of the buck-boost topology, which are used in certain applications. These two topologies are briefly described below.

3.7.2.1   SEPIC Converters (Single-Ended Primary Inductor Converters)

The SEPIC converter, shown in Figure 3.27a, is used in certain applications where the current drawn from the input is required to be relatively ripple-free. By applying Kirchhoff's voltage law and the fact that the average inductor voltage is zero in the DC steady state, the capacitor in this converter gets charged to an average value that equals the input voltage V_{in} with the polarity shown. During the on interval of the transistor, DT_s, as shown in Figure 3.27b, the diode gets reverse biased by the sum of the capacitor and the output voltages, and i_{L1} and i_{L2} flow through the transistor. During the off interval $(1-D)T_s$, i_{L1} and i_{L2} flow through the diode, as shown in Figure 3.15c. The voltage across L_2 equals v_C during the on interval and $(-V_o)$ during the off interval. In terms of the average value of the capacitor voltage that equals V_{in} (by applying Equation (3.9) in Figure 3.27a), equating the average voltage across L_2 to zero results in,

$$DV_{in} = (1-D)V_o, \tag{3.34}$$

or

$$\frac{V_o}{V_{in}} = \frac{D}{1-D}. \tag{3.35}$$

Unlike the buck-boost converters, the output voltage polarity in the SEPIC converter remains the same as that of the input.

3.7.2.2   Ćuk Converters

Named after its inventor, the Ćuk converter is shown in Figure 3.28a, where the energy transfer means is through the capacitor C between the two inductors. Using

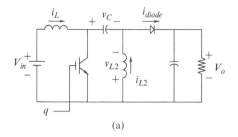

(a)

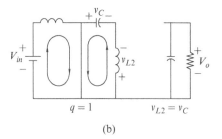

(b)

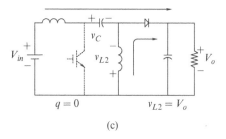

(c)

FIGURE 3.27   SEPIC converter.

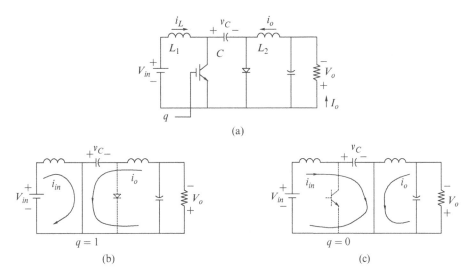

FIGURE 3.28 Ćuk converter.

Equation (3.9) in Figure 3.28a, this capacitor voltage has an average value of $(V_{in} + V_o)$ with the polarity shown. During the on interval of the transistor, DT_s, as shown in Figure 3.28b, the diode gets reverse biased by the capacitor voltage, and the input and the output currents flow through the transistor. During the off interval $(1 - D)T_s$, the input and the output currents flow through the diode, as shown in Figure 3.28c. In terms of the average values of the inductor currents, equating the net change in charge on the capacitor over T_s to zero in steady state,

$$DI_o = (1 - D)I_{in}, \tag{3.36}$$

or

$$\frac{I_{in}}{I_o} = \frac{D}{1 - D}. \tag{3.37}$$

Equating input and output powers in this idealized converter leads to,

$$\frac{V_o}{V_{in}} = \frac{D}{1 - D}, \tag{3.38}$$

which shows the same functionality as buck-boost converters. One of the advantages of the Ćuk converter is that it has non-pulsating currents at the input and the output, but it suffers from the same component stress disadvantages as the buck-boost converters and produces an output voltage of the polarity opposite to that of the input.

3.8 TOPOLOGY SELECTION [5]

For selecting between the three converter topologies discussed in this chapter, the stresses listed in Table 3.1 can be compared, which are based on the assumption that the inductor ripple current is negligible.

TABLE 3.1   Topology selection criteria.

Criterion		Buck	Boost	Buck-boost
Transistor $\hat{V}$		V_{in}	V_o	$(V_{in} + V_o)$
Transistor $\hat{I}$		I_o	I_{in}	$I_{in} + I_o$
I_{rms}	Transistor	$\sqrt{D}I_o$	$\sqrt{D}I_{in}$	$\sqrt{D}(I_{in} + I_o)$
I_{avg}	Transistor	DI_o	DI_{in}	$D(I_{in} + I_o)$
	Diode	$(1-D)I_o$	$(1-D)I_{in}$	$(1-D)(I_{in} + I_o)$
I_L		I_o	I_{in}	$I_{in} + I_o$
Effect of L on C		significant	little	little
Pulsating Current		input	output	both

From the above table, we can clearly conclude that the buck-boost converter suffers from several additional stresses. Therefore, it should be used only if both the buck and the boost capabilities are needed. Otherwise, the buck or the boost converter should be used based on the desired capability. A detailed analysis is carried out in [2].

3.9   WORST-CASE DESIGN

The worst-case design should consider the ranges in which the input voltage and the output load vary. As mentioned earlier, converters above a few tens of watts are often designed to operate in CCM. To ensure CCM even under very light load conditions would require prohibitively large inductance. Hence, the inductance value chosen is often no larger than three times the critical inductance ($L < 3L_c$), where the critical inductance L_c is the value of the inductor that will make the converter operate at the border of CCM and DCM at full load.

3.10   SYNCHRONOUS-RECTIFIED BUCK CONVERTER FOR VERY LOW OUTPUT VOLTAGES [6]

Operating voltages in computing and communication equipment have already dropped to an order of 1 V, and even lower voltages, such as 0.5 V, are predicted in the near future. At these low voltages, the diode (even a Schottky diode) of the power pole in a buck converter has an unacceptably high voltage drop across it, resulting in extremely poor converter efficiency.

As a solution to this problem, the switching power-pole in a buck converter is implemented using two MOSFETs, as shown in Figure 3.29a, which are available with very low $R_{DS(on)}$ in low voltage ratings. The two MOSFETs are driven by almost complementary gate signals (some dead time, where both signals are low, is necessary to avoid the shoot-through of current through the two transistors), as shown in Figure 3.29b.

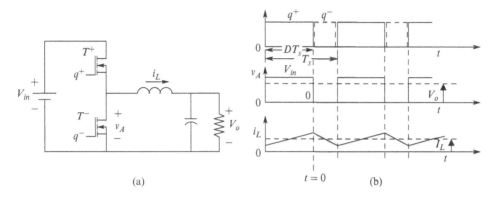

FIGURE 3.29　Buck converter: synchronous rectified.

When the upper MOSFET is off, the inductor current flows through the channel, from the source to the drain, of the lower MOSFET that has gate voltage applied to it. This results in a very low voltage drop across the lower MOSFET. At light load conditions, the inductor current may be allowed to become negative without becoming discontinuous, flowing from the drain to the source of the lower MOSFET [3].

It is possible to achieve soft switching in such converters, as discussed in Chapter 10, where the ripple in the total output and the input currents can be minimized by interleaving, which is discussed in the next section.

3.10.1　Simulation and Hardware Prototyping

The simulation of a non-ideal synchronous buck converter is demonstrated by means of an example:

Example 3.9

In the synchronous-rectified buck DC-DC converter shown in Figure 3.29a, $L = 68\,\mu H$, $C = 490\,\mu F$, and $R = 25\,\Omega$. It is operating in DC steady state under the following conditions: $V_{in} = 15\,V$, $D = 0.5$, and $f_s = 100\,kHz$. For the switch, use the parameters given in the Appendix of Chapter 2. Simulate this converter using LTspice.

Solution　The simulation file used in this example is available on the accompanying website. The LTspice model is shown in Figure 3.30, and the steady-state waveforms from the simulation of this converter are shown in Figure 3.31.

The Workbench model for implementing the above example in hardware using the Sciamble lab kit is shown in Figure 3.32.

The steady-state waveforms from running the synchronous-rectified buck converter using the Sciamble laboratory kit are shown in Figure 3.33. The step-by-step procedure for re-creating the above hardware implementation is presented in [7].

In the case of a buck converter, discussed in section 3.5, there is a significant reduction in the efficiency due to the 0.3–1.1 V drop across the freewheeling diode, as seen in Figure 3.34a. This drop can be greatly reduced in the synchronous-rectified buck converter by using a Si-MOSFET instead of a diode, as seen in Figure 3.34b.

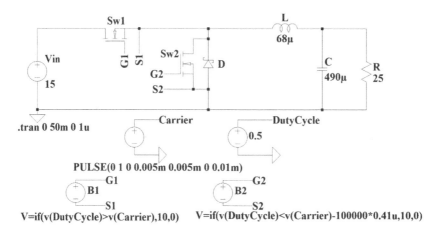

FIGURE 3.30   LTspice model.

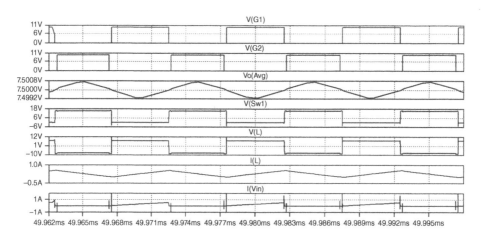

FIGURE 3.31   LTspice simulation results.

It must be noted that, while the gate ON signal transitions from one MOSFET to the other, both MOSFETs' gate is held OFF for a brief period of time to prevent shoot-through. During this short interval, the current flows through the body diode of the MOSFET, as seen by the −0.7 V drop during the dead time, of about 200 ns, in Figure 3.34b.

Figure 3.34c shows the switch-node voltage of a synchronous-rectified buck converter using GaN-FETs. The dead time required for GaN-FET of a similar rating as Si-MOSFET is typically an order of magnitude lower, 30 ns in this case. This faster switching characteristic leads to reduced switching losses. Particular care must be taken to ensure that the dead time is maintained as low as possible [8] because the voltage drop across the GaN-FET under reverse current conduction, while the gate is OFF, is typically around 3.5 V, as seen in Figure 3.34c. This reverse voltage drop is significantly higher than that of a Si-MOSFET's body diode.

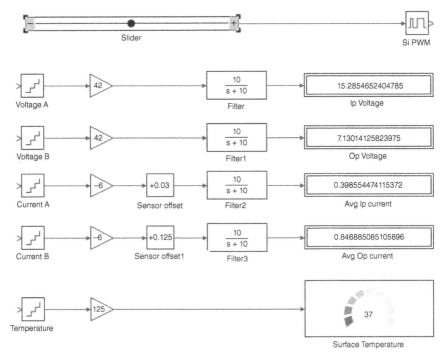

FIGURE 3.32 Workbench model.

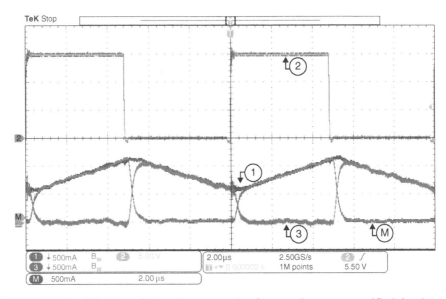

FIGURE 3.33 Workbench hardware results for synchronous-rectified buck converter using Si switches: (1) inductor current, (2) switch-node voltage, (3) high-side switch current/input current, and (M) low-side switch current.

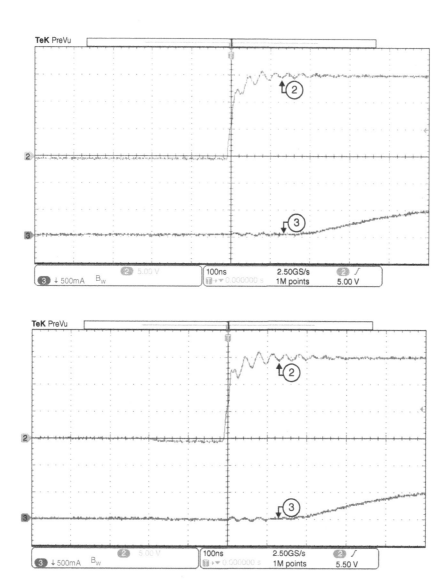

FIGURE 3.34 Switch-node voltage: (a) buck converter, (b) synchronous-rectified buck converter using Si-MOSFETs, (c) synchronous-rectified buck converter using GaN-FETs.

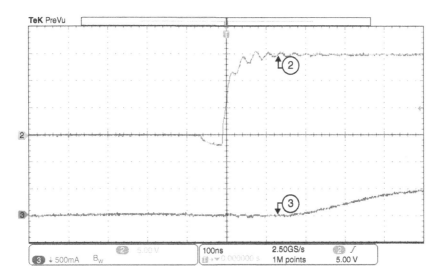

FIGURE 3.34 *Continued*

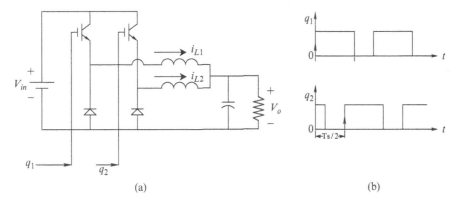

| (a) | (b) |

FIGURE 3.35 Interleaving of converters.

3.11 INTERLEAVING OF CONVERTERS

Figure 3.35a shows two interleaved converters whose switching waveforms are phase-shifted by $T_s/2$, as shown in Figure 3.35b. In general, n such converters can be used, their operation phase-shifted by T_s/n. The advantage of such interleaved multi-phase converters is the cancellation of ripple in the input and the output currents to a large degree. This is also a good way to achieve higher control bandwidth.

3.12 REGULATION OF DC-DC CONVERTERS BY PWM

Almost all DC-DC converters are operated with their output voltages regulated to equal their reference values within a specified tolerance band (for example, ±1%

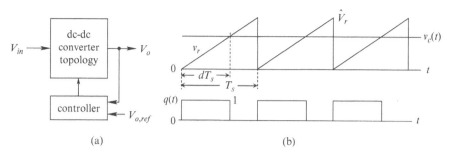

FIGURE 3.36   Regulation of output by PWM.

around its nominal value) in response to disturbances in the input voltage and the output load. The average output of the switching power-pole in a DC-DC converter can be controlled by pulsed-width modulating (PWM) the duty ratio $d(t)$ of this power pole.

Figure 3.36a shows a block diagram form of a regulated DC-DC converter. It shows that the converter output voltage is measured and compared with its reference value within a PWM controller IC, briefly described in Chapter 2. The error between the two voltages is amplified by an amplifier, whose output is the control voltage $v_c(t)$. Within the PWM-IC, the control voltage is compared with a ramp signal $v_r(t)$, as shown in Figure 3.36b, where the comparator output represents the switching function $q(t)$ whose pulse width $d(t)$ can be modulated to regulate the output of the converter.

The ramp signal v_r has the amplitude $\hat{V}_r$, and the switching frequency f_s constant. The output voltage of this comparator represents the transistor-switching function $q(t)$, which equals 1 if $v_c(t) \geq v_r$ and 0 otherwise. The switch duty ratio in Figure 3.36b is given as

$$d(t) = \frac{v_c(t)}{\hat{V}_r}, \tag{3.39}$$

and thus the control voltage, limited in a range between 0 and $\hat{V}_r$, linearly and dynamically controls the pulse width $d(t)$ in Equation (3.39) and shown in Figure 3.36b. The topic of feedback controller design for regulating the output voltage is discussed in detail in Chapter 4.

3.13   DYNAMIC AVERAGE REPRESENTATION OF CONVERTERS IN CCM

In all three types of DC-DC converters in CCM, the switching power-pole switches between two sub-circuit states based on the switching function $q(t)$. (It switches between three sub-circuit states in DCM, where the switch can be considered "stuck" between the on and the off positions during the subinterval when the inductor current is zero, discussed in detail in Section 3.15.) It is very beneficial to obtain non-switching average models of these switch-mode converters for simulating the converter performance under dynamic conditions caused by the change of input voltage and/or the output load. Under the dynamic condition, the converter duty ratio, and the average

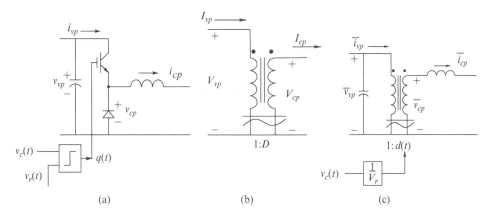

FIGURE 3.37 Average dynamic model of a switching power-pole.

values of voltages and currents within the converter vary with time, but relatively slowly, with frequencies an order of magnitude smaller than the switching frequency.

The switching power-pole is shown in Figure 3.37a, where the voltages and currents are labeled with the subscript vp for the voltage port and cp for the current port. In the above analysis for the three converters in the DC steady state, we can write the average voltage and current relationships for the bi-positional switch of the power pole as,

$$V_{cp} = DV_{vp}, \tag{3.40a}$$

$$I_{vp} = DI_{cp}. \tag{3.40b}$$

Relationships in Equation (3.40) can be represented by an ideal transformer as shown in Figure 3.37b, where the ideal transformer, hypothetical and only a convenience for mathematical representation, can operate with AC as well DC voltages and currents, which a real transformer *cannot*. A symbol consisting of a straight bar with a curve below it is used to remind us of this fact. Since no electrical isolation exists between the voltage port and the current port of the switching power-pole, the two windings of this ideal transformer in Figure 3.37b are connected at the bottom. Moreover, the voltage at the voltage port in Figure 3.37b cannot become negative, and d is limited to a range between 0 and 1.

Under dynamic conditions, the average model in Figure 3.37b of the bi-positional switch can be substituted in the power pole of Figure 3.37a, resulting in the dynamic average model shown in Figure 3.37c, using Equation (3.39) for $d(t)$. Here, the uppercase letters used in the DC steady state relationships are replaced with lowercase letters with an overbar (a "–" on top) to represent average voltages and currents, which may vary dynamically with time: D by $d(t)$, V_{cp} by $\bar{v}_{cp}(t)$, I_{cp} by $\bar{i}_{cp}(t)$, and I_{vp} by $\bar{i}_{vp}(t)$. Therefore, from Equations (3.40a) and (3.40b):

$$\bar{v}_{cp}(t) = d(t)\bar{v}_{vp}(t), \tag{3.41a}$$

$$\bar{i}_{vp}(t) = d(t)\bar{i}_{cp}(t). \tag{3.41b}$$

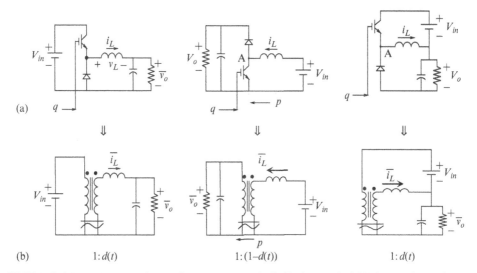

FIGURE 3.38 Average dynamic models: buck (left), boost (middle), and buck-boost (right).

The above discussion shows that the dynamic average model of a switching power-pole in CCM is an ideal transformer with the turns ratio $1{:}d(t)$. Using this model for the switching power-pole, the dynamic average models of the three converters shown in Figure 3.38a are as in Figure 3.38b in CCM. Note that in the boost converter where the transistor is in the bottom position in the power pole, the transformer turns ratio is $1{:}(1-d)$ for the following reason: Unlike buck and buck-boost converters, the transistor in a boost converter is in the bottom position. Therefore, when the transistor is on with a duty ratio d, the effective bi-positional switch is in the "down" position, and the pole duty ratio is $(1-d)$. The average representation of these converters in DCM is described in section 3.15.5.

In the average representation of the switching converters, all the switching information is removed, and hence it provides an uncluttered understanding of achieving desired objectives. Moreover, the average model in simulating the dynamic response of a converter results in computation speeds orders of magnitude faster than that in the switching model, where the simulation time-step must be smaller than at least one-hundredth of the switching time period T_s in order to achieve an accurate resolution.

3.14 BI-DIRECTIONAL SWITCHING POWER-POLE

In buck, boost, and buck-boost DC-DC converters, the implementation of the switching power-pole by one transistor and one diode dictates the instantaneous current flow to be unidirectional. As shown in Figure 3.39a, combining the switching power-pole implementations of buck and boost converters, where the two transistors are switched by complementary signals, allows a continuous bi-directional power and current capability.

In such a bi-directional switching power-pole, the positive inductor current, as shown in Figure 3.39b, represents a buck mode, where only the transistor and the diode associated with the buck converter take part. Similarly, as shown in Figure 3.39c, the negative inductor current represents a boost mode, where only the transistor and

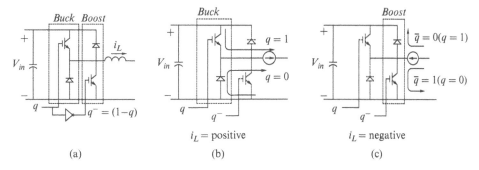

FIGURE 3.39 Bi-directional power flow through a switching power-pole.

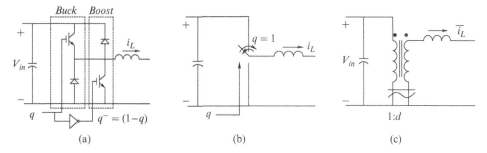

FIGURE 3.40 Average dynamic model of the switching power-pole with bi-directional power flow.

the diode associated with the boost converter take part. We will utilize such bi-directional switching power-poles in DC and AC motor drives, uninterruptible power supplies, and power systems applications, discussed in Chapters 11 and 12.

In a bi-directional switching power-pole where the transistors are gated by complementary signals, the current through it can flow in either direction, and hence ideally (ignoring a small dead time needed due to practical considerations when both the transistors are off simultaneously for a very short interval), a discontinuous conduction mode does not exist.

As described by Figures 3.39b and c above, the bi-directional switching power-pole in Figure 3.40a is in the "up" position, regardless of the direction of i_L, when $q = 1$, as shown in Figure 3.40b. Similarly, it is in the "down" position, regardless of the direction of i_L, when $q = 0$. Therefore, the average representation of the bi-directional switching power-pole in Figure 3.40a (represented as in Figure 3.40b) is an ideal transformer shown in Figure 3.40c, with a turns ratio $1:d(t)$, where $d(t)$ represents the pole duty ratio that is also the duty ratio of the transistor associated with the buck mode.

3.15 DISCONTINUOUS-CONDUCTION MODE (DCM)

All DC-DC converters for unidirectional power and current flow have their switching power-pole implemented by one transistor and one diode and hence go into a discontinuous conduction mode, DCM, below a certain output load. As an example, as shown in Figure 3.41, if we keep the switch duty ratio constant, a decline in the output load results in the inductor average current to decrease until a critical load value is reached, where the inductor current waveform reaches zero at the end of the turn-off interval.

We will call the average inductor current in this condition the critical inductor current and denote it by $I_{L,crit}$. For loads below this critical value, the inductor current cannot reverse through the diode in any of the three converters (buck, boost, and buck-boost) and enters DCM, where the inductor current remains zero for a finite interval until the transistor is turned on, beginning the next switching cycle.

In DCM, during the interval when the inductor current remains zero, there is no power drawn from the input source, and there is no energy in the inductor to transfer to the output stage. This interval of inactivity generally results in increased device stresses and the ratings of the passive components. DCM also results in noise and EMI, although the diode reverse recovery problem is minimized. Based on these considerations, converters above a few tens of watts are generally designed to operate in CCM, although all of them implemented using one transistor and one diode will enter DCM at very light loads, and the feedback controller should be designed to operate adequately in both modes. It should be noted that designing the controller of some converters, such as the buck-boost converters, in CCM is more complicated, so the designers may prefer to keep such converters in DCM for all possible operating conditions. This we will discuss further in the next chapter, dealing with the feedback controller design.

3.15.1 Critical Condition at the Border of Continuous-Discontinuous Conduction

At the critical load condition with $i_{L,crit}$ shown in Figure 3.41, the average inductor current is one-half the peak value:

$$I_{L,crit} = \frac{\hat{i}_{L,crit}}{2}. \tag{3.42}$$

At the critical condition, the peak inductor current in Figure 3.41 can be calculated by considering the voltage v_L that is applied across the inductor for an interval DT_s when the transistor is on, causing the current to rise from zero to its peak value. In buck converters, this inductor voltage is $(V_{in} - V_o)$. Using Equation (3.42) and the

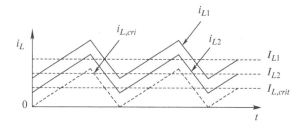

FIGURE 3.41 Inductor current at various loads; duty ratio is kept constant.

fact that $f_s = 1/T_s$ and $V_o = DV_{in}$ (the same voltage relationship holds in the critical condition as in CCM),

$$I_{L,crit,Buck} = \frac{V_{in}}{2Lf_s} D(1-D). \tag{3.43}$$

In both the boost and the buck-boost converters, the inductor voltage during DT_s equals V_{in}. Hence, using Equation (3.42),

$$I_{L,crit,Boost} = I_{L,crit,Buck-Boost} = \frac{V_{in}}{2Lf_s} D. \tag{3.44}$$

Equating the critical values of the average inductor current given in Equations (3.43) and (3.44) to input/output currents, relating the output voltage V_o to the input voltage V_{in} and recognizing that the output resistance $R = V_o/I_o$, the critical values of the load resistance in these converters can be derived as:

$$\begin{aligned}
R_{crit,Buck} &= \frac{2Lf_s}{(1-D)} \\
R_{crit,Boost} &= \frac{2Lf_s}{D(1-D)^2} \\
R_{crit,Buck-Boost} &= \frac{2Lf_s}{(1-D)^2}.
\end{aligned} \tag{3.45}$$

A load resistance above this critical value results in less than a critical load, causing the corresponding converter to go into DCM.

3.15.2 Buck Converters in DCM in Steady State

Waveforms for a buck converter under DCM are shown in Figure 3.42a. In DCM, the inductor current remains zero for a finite interval, resulting in an average value (equal to I_o) that is smaller than the critical value. When the inductor current is zero during the $D_{off,2}T_s$ interval, the voltage across the inductor is zero and $v_A = V_o$.

In a buck converter, i_{in} equals i_L during the on interval and is otherwise zero. In Figure 3.42a in DCM,

$$\hat{I}_{in} = \hat{I}_L = \frac{(V_{in} - V_o)}{L} DT_s. \tag{3.46}$$

Hence the average value of the input current, recognizing that $f_s = 1/T_s$,

$$I_{in} = \frac{(V_{in} - V_o)}{2Lf_s} D^2. \tag{3.47}$$

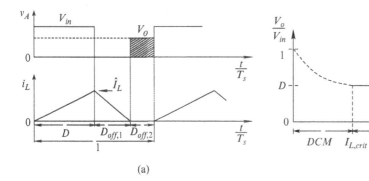

(a)   (b)

FIGURE 3.42 Buck converter in DCM.

Equating the average input power $P_{in}(=V_{in}I_{in})$ to the output power $P_o(=V_o^2/R)$, we can derive that (see problem 3.32),

$$V_o = \frac{V_{in}}{2}\left(\sqrt{M(M+4)} - M\right) \quad \text{in DCM,} \tag{3.48a}$$

where

$$M = \left(\frac{R}{2Lf_s}\right)D^2. \tag{3.48b}$$

Equation (3.48) shows that light loads with $R > R_{crit,Buck}$ cause the output voltage to rise towards the input voltage, as shown in Figure 3.42b. Of course in a regulated DC-DC converter, the feedback controller will adjust the duty ratio in order to regulate the output voltage.

3.15.2.1 Simulation and Hardware Prototyping
The simulation of a non-ideal buck converter operating under discontinuous conduction mode is demonstrated by means of an example:

Example 3.10
In the buck DC-DC converter shown in Figure 3.5a, $L = 68\,\mu H$, $C = 490\,\mu F$, and $R = 25\,\Omega$. It is operating in DC steady state under the following conditions: $V_{in} = 15\,V$, $D = 0.2$, and $f_s = 40\,kHz$. For the switch and the diode, use the parameters given in the Appendix of Chapter 2. Simulate this converter using LTspice.

Solution The simulation file used in this example is available on the accompanying website. The LTspice model is the same as the one shown in Figure 3.7, and the steady-state waveforms from the simulation of this converter are shown in Figure 3.43.

The steady-state waveforms from running the buck converter under discontinuous conduction mode using the Sciamble laboratory kit are shown in Figure 3.44. The Workbench model is the same as the one shown in Figure 3.9. The step-by-step procedure for re-creating the above hardware implementation is presented in [9].

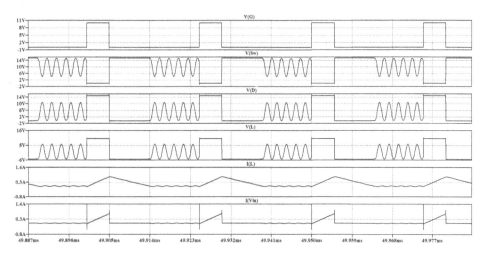

FIGURE 3.43 LTspice simulation results.

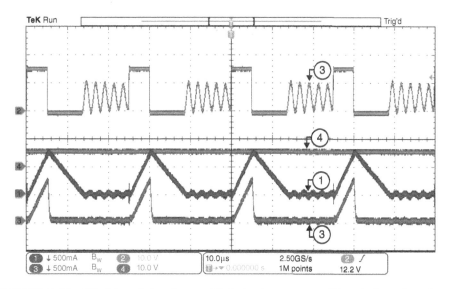

FIGURE 3.44 Workbench hardware: (1) inductor current, (2) switch-node voltage, (3) input current, and (4) output voltage.

3.15.2.2 Ringing of the Voltage at the Switching Node

Compared to the operation of the ideal buck converter in Figure 3.5a, the switch node voltage of a practical buck converter during $D_{off,2}$ rings with a peak-to-peak magnitude which is double that of the magnitude of the output voltage, clamped between 0 and V_{in}: $0 < V_A < V_{in}$. This is due to the LC tank circuit formed by the filter inductor and the parallel combination of switch and diode parasitic capacitance, as shown in Figure 3.45.

This ringing leads to increased electromagnetic interference (EMI) and higher losses. Knowing the frequency of this ringing is necessary to design an EMI filter to suppress this frequency component. The procedure is demonstrated by means of an example:

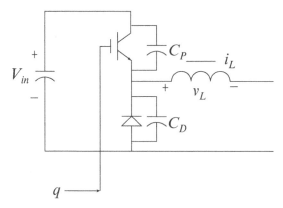

FIGURE 3.45 Non-ideal buck converter showing the parasitic switch and diode capacitance.

Example 3.11
In the buck converter shown in Figure 3.45, the switch has a parasitic capacitance of $C_T = 1550 \, pF$ and the diode has a parasitic capacitance of $C_D = 450 \, pF$. Assume the remaining parameters are the same as those given in Example 3.10. Compute the ringing frequency of the ringing of switching node voltage.

Solution $\quad f_0 = \dfrac{1}{2\pi\sqrt{L(C_T + C_D)}} = \dfrac{1}{2\pi\sqrt{68e-6(1550e-12+450e-12)}} = 0.432 \, \text{MHz}$

This matches the results shown in Figure 3.45. The magnitude and the total time period of this ringing can be greatly reduced by means of a snubber circuit, which is discussed in Chapter 8.

3.15.3 Boost Converters in DCM in Steady State

Waveforms for a boost converter under DCM are shown in Figure 3.46a. In DCM, when the inductor current is zero during the $D_{off,2}T_s$ interval, the voltage across the inductor is zero, and v_A equals V_{in}. In a boost converter, the average $V_A = V_{in}$, and i_{in} equals i_L during all intervals. In Figure 3.46a in DCM,

$$\hat{I}_{in} = \hat{I}_L = \frac{V_{in}}{L} DT_s. \tag{3.49}$$

Hence the average value of the input current, recognizing that $f_s = 1/T_s$,

$$I_{in} = \frac{V_{in}}{2Lf_s} D(D + D_{off,1}). \tag{3.50}$$

Equating the average input power $P_{in}(= V_{in}I_{in})$ to the output power $P_o(= V_o^2 / R)$,

$$\frac{V_{in}^2}{2Lf_s} D(D + D_{off,1}) = \frac{V_o^2}{R}. \tag{3.51}$$

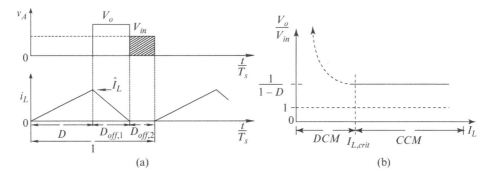

(a)   (b)

FIGURE 3.46 Boost converter in DCM.

From Figure 3.46a, the average output current, equal to V_o / R, is as follows:

$$\frac{V_{in}}{2Lf_s} DD_{off,1} = \frac{V_o}{R}.$$ (3.52)

Solving for $D_{off,1}$ from Equation (3.52) and substituting into Equation (3.51),

$$V_o = \frac{V_{in}}{2}\left(1 + \sqrt{1 + 4M}\right) \quad \text{in DCM,}$$ (3.53a)

where

$$M = \left(\frac{R}{2Lf_s}\right)D^2.$$ (3.53b)

Equation (3.53) shows that light loads with $R > R_{crit,Boost}$ cause the output voltage to rise dangerously high toward infinity, as shown in Figure 3.46b. Of course, in a regulated DC-DC converter, the feedback controller must adjust the duty ratio in order to regulate the output voltage.

3.15.3.1 Simulation and Hardware Prototyping
The simulation of a non-ideal boost converter operating under discontinuous conduction mode is demonstrated by means of an example:

Example 3.12
In the boost DC-DC converter shown in Figure 3.11a, $L = 68\,\mu H$, $C = 490\,\mu F$, and $R = 39\,\Omega$. It is operating in DC steady state under the following conditions: $V_{in} = 15\,V$, $D = 0.3$, and $f_s = 20\,kHz$. For the switch and the diode, use the parameters given in the Appendix of Chapter 2. Simulate this converter using LTspice.

Solution The simulation file used in this example is available on the accompanying website. The LTspice model is the same as the one shown in Figure 3.15, and the steady-state waveforms from the simulation of this converter are shown in Figure 3.47.

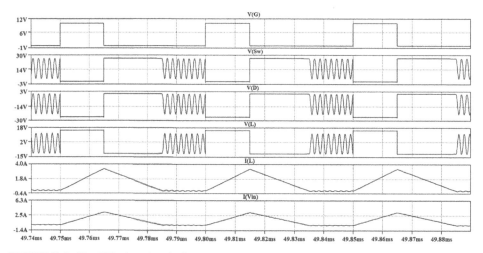

FIGURE 3.47   LTspice simulation results.

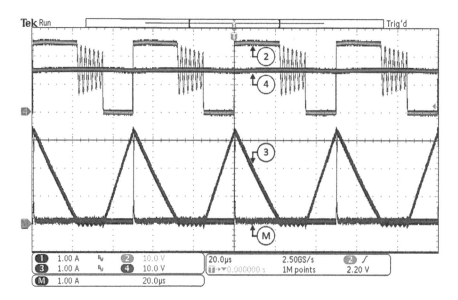

FIGURE 3.48   Workbench hardware: (2) switch-node voltage, (3) diode current, (4) output voltage, and (M) switch current.

The steady-state waveforms from running the boost converter under discontinuous conduction mode using the Sciamble laboratory kit are shown in Figure 3.48. The Workbench model is the same as the one shown in Figure 3.17. The step-by-step procedure for re-creating the above hardware implementation is presented in [10].

Similar to the practical buck converter, the switch-node voltage of the practical boost converter also rings, during $D_{off,2}$ with a peak-to-peak magnitude which is double that of the magnitude of the difference between the output and the input voltage, clamped between 0 and V_0: $0 < V_A < V_o$.

3.15.4 Buck-Boost Converters in DCM in Steady State

Waveforms for a buck-boost converter under DCM are shown in Figure 3.49a. In DCM, when the inductor current is zero during $D_{off,2}T_s$ interval, the voltage across the inductor is zero and v_A equals V_o. In a buck-boost converter, the average $V_A = V_o$.

In a buck-boost converter, i_{in} equals i_L during the on interval and is otherwise zero. In Figure 3.49a in DCM,

$$\hat{I}_{in} = \hat{I}_L = \frac{V_{in}}{L} DT_s. \tag{3.54}$$

Hence, the average value of the input current, recognizing that $f_s = 1/T_s$,

$$I_{in} = \frac{V_{in}}{2Lf_s} D^2. \tag{3.55}$$

Equating the average input power $P_{in}(= V_{in}I_{in})$ to the output power $P_o(= V_o^2/R)$,

$$\frac{V_{in}^2}{2Lf_s} D^2 = \frac{V_o^2}{R}. \tag{3.56}$$

Therefore,

$$V_o = DV_{in}\sqrt{\frac{R}{2Lf_s}} \quad \text{in DCM.} \tag{3.57}$$

Equation (3.57) shows that light loads with $R > R_{crit,Buck-Boost}$ cause the output voltage to rise dangerously high toward infinity, as shown in Figure 3.49b. Of course, in a regulated DC-DC converter, the feedback controller must adjust the duty ratio in order to regulate the output voltage.

3.15.4.1 Simulation and Hardware Prototyping

The simulation of a non-ideal buck-boost converter operating under discontinuous conduction mode is demonstrated by means of an example:

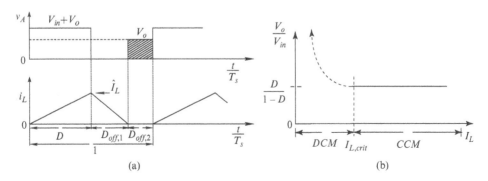

(a) (b)

FIGURE 3.49 Buck-boost converter in DCM.

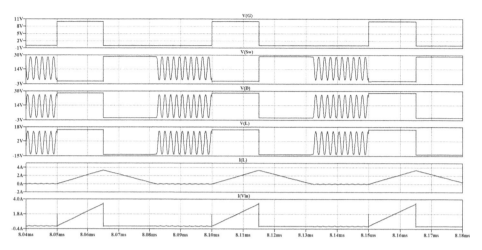

FIGURE 3.50   LTspice simulation results.

Example 3.13
In the buck-boost DC-DC converter shown in Figure 3.19b, $L = 68 \mu H$, $C = 490 \mu F$, and $R = 39 \Omega$. It is operating in DC steady state under the following conditions: $V_{in} = 15$ V, $D = 0.3$, and $f_s = 20$ kHz. For the switch and the diode, use the parameters given in the Appendix of Chapter 2. Simulate this converter using LTspice.

Solution   The simulation file used in this example is available on the accompanying website. The LTspice model is the same as the one shown in Figure 3.23, and the steady-state waveforms from the simulation of this converter are shown in Figure 3.50.

The steady-state waveforms from running the buck-boost converter under discontinuous conduction mode using the Sciamble laboratory kit are shown in Figure 3.51. The Workbench model is the same as the one shown in Figure 3.25. The step-by-step procedure for re-creating the above hardware implementation is presented in [11].

Similar to the practical buck converter, the switch-node voltage of the practical buck-boost converter also rings, during $D_{off,2}$ with a peak-to-peak magnitude which is double that of the magnitude of the output voltage, clamped between 0 and $V_{in} + V_o$: $0 < V_A < V_{in} + V_o$.

3.15.5   Average Representation in CCM and DCM for Dynamic Analysis

In the previous sections, all three DC-DC converters are analyzed in DCM in steady state. Unlike the CCM, where the output voltage is dictated only by the input voltage V_{in} and the transistor duty ratio, the output voltage in DCM also depends on the converter parameters and the operating condition. The output voltage at the voltage port of the switching power-pole is higher than that in the CCM case.

If the duty ratio varies slowly, with a frequency an order of magnitude smaller than the switching frequency, then the average representation obtained on the basis of the DC steady state can be used for dynamic modeling in CCM and DCM by replacing uppercase letters with lowercase letters with an overbar (a "–" on top) to represent average quantities that are shown explicitly to be functions of time:

$$D \rightarrow d(t) \quad V_o \rightarrow \overline{v}_o(t) \quad I_L \rightarrow \overline{i}_L(t) \quad I_{in} \rightarrow \overline{i}_{in}. \tag{3.58}$$

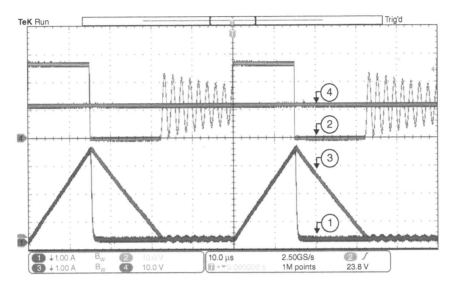

FIGURE 3.51 Workbench hardware: (1) input current, (2) switch-node voltage, (3) inductor current, and (4) output voltage.

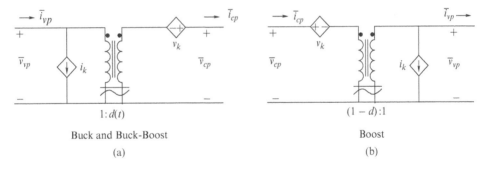

FIGURE 3.52 Average representation of a switching power-pole valid in CCM and DCM.

Therefore, as derived in the Appendix, in DCM, the average model of a switching power-pole in CCM by an ideal transformer is augmented by a dependent voltage source v_k at the current port and by a dependent current source i_k at the voltage port, as shown in Figure 3.52a for buck and buck-boost converters and in Figure 3.52b for boost converters. The values of these dependent sources for the three converters are calculated in the Appendix, and only the results are presented in Table 3.2.

If v_k and i_k are expressed conditionally such that they are both zero in CCM and are expressed by the expressions in Table 3.2 only in DCM when the average inductor current falls below its critical value, then the representations in Figure 3.52 become valid for both the CCM and the DCM. In LTspice the ideal transformer itself is represented by a dependent current source and a dependent voltage source. The conditional dependent sources v_k and i_k shown in Figures 3.52a and 3.52b are in addition to those used to represent the ideal transformer.

TABLE 3.2 V_k and i_k.

Converter	v_k	i_k
Buck	$(1-\dfrac{2Lf_s\bar{i}_L}{(V_{in}-\bar{v}_o)d})\bar{v}_o$	$\dfrac{d^2}{2Lf_s}(V_{in}-\bar{v}_0)-d\bar{i}_L$
Boost	$(1-\dfrac{2Lf_s\bar{i}_L}{V_{in}d})(V_{in}-\bar{v}_0)$	$\dfrac{d^2}{2Lf_s}V_{in}-d\bar{i}_L$
Buck-Boost	$(1-\dfrac{2Lf_s\bar{i}_L}{V_{in}d})\bar{v}_o$	$\dfrac{d^2}{2Lf_s}V_{in}-d\bar{i}_L$

REFERENCES

1. N. Mohan, T.M. Undeland, and W.P. Robbins, *Power Electronics: Converters, Applications and Design*, 3rd Edition (New York: John Wiley & Sons, 2003).
2. Buck Converter Lab Manual. https://sciamble.com/resources/pe-drives-lab/basic-pe/buck-converter.
3. Boost Converter Lab Manual. https://sciamble.com/resources/pe-drives-lab/basic-pe/boost-converter.
4. Buck-Boost Converter Lab Manual. https://sciamble.com/resources/pe-drives-lab/basic-pe/buck-boost-converter.
5. B. Carsten, "Converter Component Load Factors: A Performance Limitation of Various Topologies," *PCI Proceedings*, June 1988, pp. 31–49.
6. M. Walters, "An Integrated Synchronous-Rectifier Power IC with Complementary-Switching (HIP5010, HIP5011)," Technical Brief, July 1995, TB332, Intersil Corp.
7. Synchronous-Rectified Buck Converter Lab Manual. https://sciamble.com/resources/pe-drives-lab/basic-pe/synchronous-buck-converter.
8. "Does GaN Have a Body Diode?—Understanding the Third Quadrant Operation of GaN," Texas Instruments Application Report SNOAA36–February 2019. https://www.ti.com/lit/an/snoaa36/snoaa36.pdf?ts=1654535176366.
9. Buck Converter Discontinuous Conduction Mode Lab Manual. https://sciamble.com/resources/pe-drives-lab/basic-pe/buck-converter-dcm.
10. Boost Converter Discontinuous Conduction Mode Lab Manual. https://sciamble.com/resources/pe-drives-lab/basic-pe/boost-converter-dcm.
11. Buck-Boost Converter Discontinuous Conduction Mode Lab Manual. https://sciamble.com/resources/pe-drives-lab/basic-pe/buck-boost-converter-dcm.

PROBLEMS

Buck DC-DC Converters

Problems 3.1 through 3.6: In a buck DC-DC converter, $L= 25\,\mu H$. It is operating in DC steady state under the following conditions: $V_{in} = 42\,V$, $D = 0.3$, $P_o = 24\,W$, and $f_s = 400\,kHz$. Assume ideal components.

3.1   Calculate and draw the waveforms as shown in Figure 3.5d.
3.2   Draw the inductor voltage and current waveforms if $P_o = 12\,W$; all else is unchanged. Compare the ripple in the inductor current with that in Problem 3.1.

3.3   In this buck converter, the output load is changing. Calculate the critical value of the output load P_o below which the converter will enter the discontinuous conduction mode of operation.

3.4   Calculate the critical value of the inductance L such that this buck converter remains in the continuous conduction mode at and above $P_o = 5\,\text{W}$ under all values of the input voltage V_{in} in a range from 24 V to 50 V.

3.5   Draw the waveforms for the variables shown in Figure 3.5d for this buck converter at the output load that causes it to operate at the border of continuous and discontinuous modes.

3.6   In this buck converter, the input voltage is varying in a range from 24 V to 50 V. For each input value, the duty ratio is adjusted to keep the output voltage constant at its nominal value (with $V_{in} = 40\,\text{V}$ and $D = 0.3$). Calculate the minimum value of the inductance L that will keep the converter in the continuous conduction mode at $P_o = 5\,\text{W}$.

3.7   A buck DC-DC converter is to be designed for $V_{in} = 20\,\text{V}$, $V_o = 12\,\text{V}$, and the maximum output power $P_o = 72\,\text{W}$. The switching frequency is selected to be $f_s = 400\,\text{kHz}$. Assume ideal components. Estimate the value of the filter inductance that should be used if the converter is to remain in CCM at one-third the maximum output power.

Boost DC-DC Converters

Problems 3.8 through 3.13: In a boost converter, $L = 25\,\mu\text{H}$. It is operating in DC steady state under the following conditions: $V_{in} = 12\,\text{V}$, $D = 0.4$, $P_o = 25\,\text{W}$, and $f_s = 400\,\text{kHz}$. Assume ideal components.

3.8   Calculate and draw the waveforms as shown in Figure 3.12c.

3.9   Draw the inductor voltage and current waveforms, if $P_o = 15\,\text{W}$; all else is unchanged. Compare the ripple in the inductor current with that in Problem 3.8.

3.10 In this boost converter, the output load is changing. Calculate the critical value of the output load P_o below which the converter will enter the discontinuous conduction mode of operation.

3.11 Calculate the critical value of the inductance L below which this boost converter will enter the discontinuous conduction mode of operation at $P_o = 5\,\text{W}$.

3.12 Draw the waveforms for the variables in Figure 3.12c for this boost converter at the output load that causes it to operate at the border of continuous and discontinuous modes.

3.13 In this boost converter, the input voltage is varying in a range from 9 V to 15 V. For each input value, the duty ratio is adjusted to keep the output voltage constant at its nominal value (with $V_{in} = 12\,\text{V}$ and $D = 0.4$). Calculate the critical value of the inductance L such that this boost converter remains in the continuous conduction mode at and above $P_o = 5\,\text{W}$ under all values of the input voltage V_{in}.

3.14 A boost converter is to be designed with the following values: $V_{in} = 5\,\text{V}$, $V_o = 12\,\text{V}$, and the maximum output power $P_o = 40\,\text{W}$. The switching frequency is selected to be $f_s = 400\,\text{kHz}$. Assume ideal components. Estimate the value of L if the converter is to remain in CCM at one-third the maximum output power.

Boost DC-DC Converters

Problems 3.15 through 3.20: In a buck-boost converter, $L = 25\,\mu H$. It is operating in DC steady state under the following conditions: $V_{in} = 12\,V$, $D = 0.6$, $P_o = 36\,W$, and $f_s = 400\,kHz$. Assume ideal components.

3.15   Calculate and draw the waveforms as shown in Figure 3.20c.

3.16   Draw the inductor voltage and current waveforms if $P_o = 18\,W$; all else is unchanged. Compare the ripple in the inductor current with that in Problem 3.15.

3.17   In this buck-boost converter, the output load is changing. Calculate the critical value of the output load P_o below which the converter will enter the discontinuous conduction mode of operation.

3.18   Calculate the critical value of the inductance L below which this buck-boost converter will enter the discontinuous conduction mode of operation at $P_o = 5\,W$.

3.19   Draw the waveforms for the variables in Figure 3.20c for this buck-boost converter at the output load that causes it to operate at the border of continuous and discontinuous modes.

3.20   In this buck-boost converter, the input voltage is varying in a range from 9 V to 15 V. For each input value, the duty ratio is adjusted to keep the output voltage constant at its nominal value (with $V_{in} = 12\,V$ and $D = 0.6$). Calculate the critical value of the inductance L such that this buck-boost converter remains in the continuous conduction mode at and above $P_o = 5\,W$ under all values of the input voltage V_{in}.

3.21   A buck-boost converter is to be designed under the following conditions: $V_{in} = 12\,V$, $V_o = 42\,V$, and the maximum output power is $P_o = 96\,W$. The switching frequency is selected to be $f_s = 300\,kHz$. Assume ideal components. Estimate the value of L if the converter is to remain in CCM at one-third the maximum output power.

SEPIC DC-DC Converters

3.22   In a SEPIC converter, assume the ripple in the inductor currents and the capacitor voltage to be zero. This SEPIC converter is operating in a DC steady state under the following conditions:$V_{in} = 10\,V$, $D = 0.333$, $P_o = 50\,W$, and $f_s = 200\,kHz$. Assume ideal components. Draw the waveforms for all the converter variables under this DC steady-state condition.

3.23   Show that the SEPIC converter consists of a switching power-pole as its building block.

Ćuk DC-DC Converters

3.24   In a Ćuk converter, assume the ripple in the inductor currents and the capacitor voltage to be zero. This Ćuk converter is operating in a DC steady state under the following conditions: $V_{in} = 10\,V$, $D = 0.333$, $P_o = 50\,W$, and $f_s = 200\,kHz$. Assume ideal components. Draw the waveforms for all the converter variables under this DC steady-state condition.

3.25 Show that the Ćuk converter consists of a switching power-pole as its building block.

Interleaving of DC-DC Converters

3.26 Two interleaved buck converters, each similar to the buck converter for Problem 3.1, are supplying a total of 48 W. Calculate and draw the waveforms of the total input current and the total current ($i_{L_1} + i_{L_2}$) to the output stage. The gate signals to the converters are phase shifted by $180°$.

Regulation by PWM

3.27 In a buck converter, consider two values of duty ratios: 0.3 and 0.4. The switching frequency f_s is 200 kHz and $\hat{V}_r = 1.2$ V. Draw the waveforms as in Figure 3.36b.

Dynamic Average Models in CCM

3.28 Draw the dynamic average representations for buck, boost, buck-boost, SEPIC, and Ćuk converters in the continuous conduction mode for representation in LTspice.

3.29 In the converters based on average representations in Problem 3.28, calculate the average input current for each converter in terms of the average output current and the duty ratio $d(t)$.

Bi-Directional Switching Power-Poles

3.30 The DC-DC bi-directional converter shown in Figure 3.40a interfaces a 12/14-V battery with a 36/42-V battery bank. The internal emfs are $E_1 = 40$ V (DC) and $E_2 = 13$ V (DC). Both these battery sources have an internal resistance of 0.1Ω each. In the DC steady state, calculate the power pole duty ratio D_A if (a) the power into the low-voltage battery terminals is 140 W, and (b) the power out of the low-voltage battery terminals is 140 W.

DC-DC Converters in DCM

3.31 Derive the expressions for the critical resistance for the three converters given in Equation (3.45).

3.32 Derive the voltage ratio in DCM for a buck converter, given by Equation (3.48).

3.33 Derive the voltage ratio in DCM for a boost converter, given by Equation (3.53).

3.34 Derive the voltage ratio in DCM for a buck-boost converter, given by Equation (3.57).

3.35 Repeat the buck converter Problem 3.1 if P_o is not specified but the load resistance is twice its critical value.

3.36 Repeat the boost converter Problem 3.8 if P_o is not specified but the load resistance is twice its critical value.

3.37 Repeat the buck-boost converter Problem 3.15 if P_o is not specified, but the load resistance is twice its critical value.

Sustainability-Related Questions

3.38 A 1.2 kW PV system consists of 4 series-connected arrays, each of which has a rated voltage of 55 V and a rated current of 6.0 A at the maximum power point, at the maximum expected insolation. The output of the series-connected arrays is boosted to 400 V (DC) by a boost converter as an input to an inverter that feeds power into the single-phase utility grid. Assuming that the switching frequency in the boost converter is selected to be 200 kHz and the peak-to-peak ripple in the input current to this converter is to be less than 10% of the rated current at the maximum power, calculate the value of the inductance in this converter. Assume ideal components.

3.39 In a large battery-storage system, the rated output voltage of a battery string is 650 V (DC), and the rated current is 1,200 A. The voltage can vary in a range from 500 V to 800 V. This output voltage is boosted to 900 V (DC), as an input to an inverter that interfaces with the three-phase utility grid. A switching power-pole with a bi-directional power capability is used. The switching frequency is selected to be 50 kHz, and the inductor value is selected such that the peak-to-peak ripple in the inductor current is less than 10% of the rated battery current, with the battery voltage at its rated value.
   (a) Calculate the inductance value.
   (b) Calculate the peak-to-peak ripple in the inductor current if operating at the rated current, but the battery voltage is at its minimum.
   (c) Repeat part (b) if the battery voltage is at its maximum.

Simulation Problems

3.40 In the buck converter shown in Figure 3.5a, various parameters are as follows: $L = 100\,\mu H$, $R_L = 10\,m\Omega$, $C = 100\,\mu F$ and $R_{Load} = 9.0\,\Omega$. The input voltage $V_{in} = 24\,V$, the switching frequency $f_s = 100\,kHz$, and the switch duty ratio $d = 0.75$.
   (a) Plot the waveforms, after reaching the steady state, during the last 10 switching cycles for i_L, v_L, and v_o.
   (b) Plot the average value of v_L.
   (c) Plot i_L and measure the peak-peak ripple Δi_L and compare it with Equation (3.13).
   (d) Plot the i_C waveform. What is the average of i_C? Compare the i_C waveform with the ripple in i_L.
   (e) Plot the input current waveform and calculate its average. Compare it to the value calculated from Equation (3.16).

3.41 In Problem 3.40, calculate the inductance value of L, if Δi_L should be 1/3 of the load current. Verify the results by simulations.

3.42 In Problem 3.40, change the output power to one-half its original value. Measure the peak-peak ripple Δi_L and compare it with that in Problem 3.40c. Comment on this comparison.

3.43 In the circuit of Problem 3.40, calculate R_{crit} from Equation (3.45) and verify by simulation whether the converter is operating on the boundary of CCM and DCM.

3.44   In the boost converter shown in Figure 3.11, various parameters are as follows: $L = 100\,\mu H$, $R_L = 10\,m\Omega$, $C = 100\,\mu F$ and $R_{Load} = 12.5\,\Omega$. The input voltage $V_{in} = 10\,V$, the switching frequency $f_s = 100\,kHz$, and the switch duty ratio $d = 0.6$.

   (a)   Plot the waveforms during the last 10 switching cycles, after reaching the steady state, for i_L, v_L, and v_o.

   (b)   Plot the average value of v_L.

   (c)   Plot i_L and measure the peak-peak ripple Δi_L and compare it with Equation (3.21).

   (d)   Plot the i_C and i_{diode} waveform. What is the average of i_C? Compare the i_C waveform with the ripple in i_{diode}.

   (e)   Plot the input current waveform and calculate its average. Compare it to the value calculated from Equation (3.23).

3.45   In Problem 3.44, calculate the inductance value of L, if Δi_L should be 1/3 of the input current. Verify the results by simulations.

3.46   In Problem 3.44, change the output power to one-half its original value. Measure the peak-peak ripple Δi_L and compare it with that in Problem 3.44c. Comment on this comparison.

3.47   In Problem 3.44, calculate R_{crit} from Equation (3.45) and verify whether the converter is operating on the boundary of CCM and DCM.

3.48   In the buck-boost converter shown in Figure 3.11, various parameters are as follows: $L = 100\,\mu H$, $R_L = 10\,m\Omega$, $C = 100\,\mu F$ and $R_{Load} = 15.0\,\Omega$. The input voltage $V_{in} = 10\,V$, the switching frequency $f_s = 100\,kHz$, and the switch duty ratio $d = 0.75$.

   (a)   Plot the waveforms during the last 10 switching cycles, after reaching the steady state, for i_L, v_L and v_o.

   (b)   Plot the average value of v_L.

   (c)   Plot i_L and measure the peak-peak ripple Δi_L and compare it with Equation (3.28).

   (d)   Plot the i_C and i_{diode} waveform. What is the average of i_C? Compare the i_C waveform with the ripple in i_{diode}.

   (e)   Plot the input current waveform and calculate its average. Compare it to the value calculated from Equation (3.31).

3.49   In Problem 3.48, calculate the inductance value of L, if Δi_L should be 1/3 of the input current. Verify the results by simulations.

3.50   In Problem 3.48, change the output power to one-half its original value. Measure the peak-peak ripple Δi_L and compare it with that in Problem 3.48c. Comment on this comparison.

3.51   In Problem 3.48, calculate R_{crit} from Equation (3.45) and verify whether the converter is operating on the boundary of CCM and DCM.

3.52   The buck-boost converter of Problem 3.48 is initially operating with an output load resistance $R_{Load} = 30.0\,\Omega$. After the steady state is reached, an additional resistance of $30.0\,\Omega$ is switched in. Simulate this converter by representing the switching power-pole by its average model and compare the waveform of the inductor current in the average representation to that in switching representation.

3.53   The buck-boost converter in Problem 3.48 is initially operating in DCM with $R_{Load} = 1\,k\Omega$.

(a)   Plot the waveforms during the last 5 switching cycles for i_L, v_L, and v_o.
(b)   Plot the average value of v_L.
(c)   Calculate V_o using Equation (3.57) and compare it with its measured value.
(d)   Plot the waveform of v_A, label it in terms of V_{in} and V_o, and compare it with Figure 3.49a.
(e)   Show that the hatched area in Figure 3.49a, averaged over the switching time period, results in the increase in V_o, compared to its CCM value.

APPENDIX 3A AVERAGE REPRESENTATION IN DISCONTINUOUS-CONDUCTION MODE (DCM)

3A.1   Introduction

Single-switch converters such as buck, boost, and buck-boost DC converters enter the discontinuous-conduction mode (DCM) under light-load conditions. In DCM, the inductor current becomes zero for a finite interval during a switching cycle. In the continuous-conduction mode (CCM), the building-block switch has two positions, either up or down. In DCM, the switch has another position (in the middle), thus resulting in three circuit sub-states during a switching cycle.

The following analysis shows that the average representation of single-switch converters in DCM consists of an ideal transformer (similar to that in CCM), augmented by dependent voltage and current sources. Such a representation allows a smooth transition between continuous and discontinuous modes.

3A.2   Average Representation in Discontinuous-Conduction Mode

3A.2.1   Buck and Buck-Boost Converters

The basic building block, with the switch and the diode in the appropriate positions for the buck and the buck-boost converters, is shown in Figure 3A.1a. The currents i_{cp} and i_{vp} and the voltage v_{cp} are shown in Figure 3A.1b, where v_{L2} is the voltage at the second terminal of the inductor, as defined in Figure 3A.1a. The switch duty ratio, controlled by the PWM-IC, is defined as d. The inductor current flows through the diode during $d_{off,1}$, and the inductor current remains zero during $d_{off,2}$.

Based on voltages in the circuit of Figure 3A.1a, and the waveforms in Figure 3A.1b,

$$i_{pk} = \frac{V_{vp} - v_{L2}}{L} dT_s = \frac{v_{L2}}{L} d_{off,1}T_s, \tag{3A.1}$$

and

$$\bar{i}_{cp} = \frac{d + d_{off,1}}{2} i_{pk}. \tag{3A.2}$$

From Equations 3A.1 and 3A.2,

$$d + d_{off,1} = \frac{2\bar{i}_{cp}L}{(V_{vp} - v_{L2})dT_s} = \frac{k\bar{i}_{cp}}{(V_{vp} - v_{L2})d}, \tag{3A.3}$$

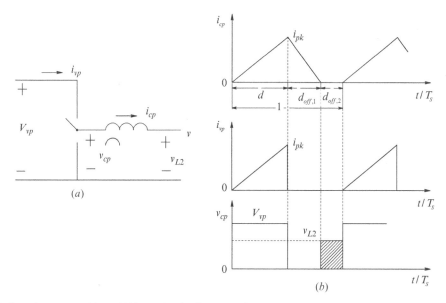

FIGURE 3A.1   (a) Building Block. (b) Waveforms.

where

$$k = 2Lf_s \tag{3A.4}$$

is a constant in terms of the inductance and the switching frequency $f_s = 1/T_s$.
The average equivalent circuit is shown in Figure 3.28a where the ideal transformer
with the turns ratio $1:d$ is augmented by two dependent sources i_k and v_k. The expressions for these dependent sources can be calculated easily, as shown below:

In the model of Figure 3.52a,

$$i_k = \bar{i}_{vp} - d\bar{i}_{cp}. \tag{3A.5}$$

From the waveforms in Figure 3A.1b,

$$\bar{i}_{vp} = \frac{d}{d+d_{off,1}} \bar{i}_{cp}. \tag{3A.6}$$

Substituting for $\bar{i}_{vp}$ from Equation 3A.6, and for $(d+d_{off,1})$ from Equation 3A.3 into
Equation 3A.5,

$$i_k = \frac{d^2}{k}(V_{vp} - v_{L2}) - d\bar{i}_{cp}. \tag{3A.7}$$

To calculate v_k in Figure 3.28a, the average of the voltage at the current port is

$$\bar{v}_{cp} = V_{vp}d + 0 \cdot d_{off,1} + d_{off,2}v_{L2} = V_{vp}d + \left[1-(d+d_{off,1})\right]v_{L2}. \tag{3A.8}$$

From Figure 3A.2,

$$v_k = \overline{v}_{cp} - dV_{vp}.$$
(3A.9)

Substituting Equation 3A.8 and Equation 3A.3 into Equation 3A.9,

$$v_k = \left[1 - \frac{k\overline{i}_{cp}}{(V_{vp} - v_{L2})d}\right] v_{L2}.$$
(3A.10)

At the transition from the DCM to CCM, both i_k and v_k must go to zero simultaneously in the model of Figure 3.52a. The critical value of the current at the boundary of DCM and CCM, from either Equation 3A.7 for i_k or Equation 3A.10 for v_k is

$$\overline{i}_{cp,crit} = \frac{d}{k}(V_{vp} - v_{L2}).$$
(3A.11)

3A.2.1.1 Expressions for Buck Converters

The above analysis was general, applicable to buck as well as buck-boost converters. Specifically in buck converters, $V_{vp} = V_{in}$, $v_{L2} = \overline{v}_0$, and $\overline{i}_{cp} = \overline{i}_L$. Therefore,

$$i_{k,buck} = \frac{d^2}{2Lf_s}(V_{in} - \overline{v}_0) - d\overline{i}_L.$$
(3A.12a)

$$v_{k,buck} = \left[1 - \frac{2Lf_s\overline{i}_L}{(V_{in} - \overline{v}_0)d}\right]\overline{v}_0.$$
(3A.12b)

and

$$\overline{i}_{L,crit,buck} = \frac{d}{2Lf_s}(V_{in} - \overline{v}_0).$$
(3A.12c)

3A.2.1.2 Expressions for Buck-Boost Converters

In buck-boost converters, $V_{vp} = V_{in} + \overline{v}_0$, $v_{L2} = \overline{v}_0$, and $\overline{i}_{cp} = \overline{i}_L$. Therefore,

$$i_{k,buck-boost} = \frac{d^2}{2Lf_s}V_{in} - d\overline{i}_L,$$
(3A.13a)

$$v_{k,buck-boost} = \left[1 - \frac{2Lf_s\overline{i}_L}{V_{in}d}\right]V_0,$$
(3A.13b)

and

$$\overline{i}_{L,crit,buck-boost} = \frac{d}{2Lf_s}V_{in}.$$
(3A.13c)

3A.2.2 Boost Converters

The average representation for boost converters operating in DCM is shown in Figure 3.52b. A similar analysis as for buck and buck-boost converters results in the following equations:

$$i_{k,boost} = \frac{d^2}{2Lf_s}V_{in} - d\bar{i}_L,$$
(3A.14a)

$$v_{k,boost} = \left[1 - \frac{2Lf_s\bar{i}_L}{V_{in}d}\right](V_{in} - \bar{v}_0),$$
(3A.14b)

and

$$\bar{i}_{L,crit,boost} = \frac{d}{2Lf_s}V_{in}.$$
(3A.14c)

3A.3   Average Modeling in PSpice for Large Signal Disturbances

To keep it general so that the models in PSpice are valid both in CCM and DCM, the dependent sources in Figure 3.52a and Figure 3.52b should be represented by conditional expressions, which make them go to zero in CCM.

4

DESIGNING FEEDBACK CONTROLLERS IN SWITCH-MODE DC POWER SUPPLIES

In most applications discussed in Chapter 1, power electronic converters are operated in a controlled manner. The need for doing so is evident in electric drives used in transportation to control speed and position. The same is also true in photovoltaic systems, where we should operate at their maximum power point to derive the maximum power. In wind turbines, the generator speed should be controlled to operate the turbine blades at the maximum value of the turbine coefficient of performance. In DC-DC converters, with or without electrical isolation, the output voltage needs to be regulated at a specified value with a narrow tolerance. In this chapter, the fundamental concepts for feedback control are illustrated by means of regulated DC-DC converters.

4.1 INTRODUCTION AND OBJECTIVES OF FEEDBACK CONTROL

As shown in Figure 4.1, almost all DC-DC converters operate with their output voltage regulated to equal their reference value within a specified tolerance band (for example, $\pm 1\%$ around its nominal value) in response to disturbances in the input voltage and the output load. This regulation is achieved by pulsed-width-modulating the duty ratio $d(t)$ of their switching power-pole. In this chapter, we will design the feedback controller to regulate the output voltages of DC-DC converters.

The feedback controller to regulate the output voltage must be designed with the following objectives in mind: zero steady-state error, fast response to changes in the input voltage and the output load, low overshoot, and low noise susceptibility. We should note that in designing feedback controllers, all transformer-isolated topologies discussed later in Chapter 8 can be replaced by their basic single-switch topologies from which they are derived. The feedback control is described using the voltage-mode control, which is later extended to include the current-mode control.

Power Electronics A First Course: Simulations and Laboratory Implementations, Second Edition.
Ned Mohan and Siddharth Raju.
© 2023 John Wiley & Sons, Inc. Published 2023 by John Wiley & Sons, Inc.
Companion Website: www.wiley.com/go/mohan/powerelectronics2e

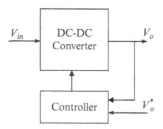

FIGURE 4.1   Regulated DC power supply.

The steps in designing the feedback controller are described as follows:

- Linearize the system for small changes around the DC steady-state operating point (bias point). This requires dynamic averaging, discussed in the previous chapter.
- Design the feedback controller using linear control theory.
- Confirm and evaluate the system response by simulations for large disturbances.

4.2   REVIEW OF LINEAR CONTROL THEORY

A feedback control system is shown in Figure 4.2, where the output voltage is measured and compared with a reference value V_o^*. The error v_{err} between the two acts on the controller, which produces the control voltage $v_c(t)$. This control voltage acts as the input to the pulse-width modulator to produce a switching signal $q(t)$ for the power pole in the DC-DC converter. The average value of this switching signal is $d(t)$, as shown in Figure 4.2.

To make use of linear control theory, various blocks in the power supply system of Figure 4.2 are linearized around the steady-state DC operating point, assuming small-signal perturbations. Each average quantity (represented by an overbar, i.e. a "–" on top) associated with the power pole of the converter topology can be expressed as the sum of its steady-state DC value (represented by an uppercase letter) and a small-signal perturbation (represented by a "~" on top), for example,

$$\bar{v}_o(t) = V_o + \tilde{v}_0(t)$$
$$d(t) = D + \tilde{d}(t) \tag{4.1}$$
$$v_c(t) = V_c + \tilde{v}_c(t),$$

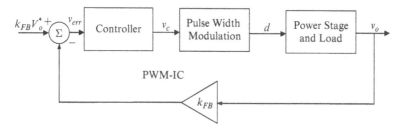

FIGURE 4.2   Feedback control.

where $d(t)$ is already an averaged value and $v_c(t)$ does not contain any switching frequency component. Based on the small-signal perturbation quantities in the Laplace domain, the linearized system block diagram is as shown in Figure 4.3, where the perturbation in the reference input to this feedback-controlled system, $\tilde{v}_0^*$, is zero since the output voltage is being regulated to its reference value. In Figure 4.3, $G_{PWM}(s)$ is the transfer function of the pulse-width modulator and $G_{PS}(s)$ is the power stage transfer function. In the feedback path, the transfer function is of the voltage-sensing network, which can be represented by a simple gain k_{FB}, usually less than unity. $G_C(s)$ is the transfer function of the feedback controller that needs to be determined to satisfy the control objectives.

As a review, the Bode plots of transfer functions with poles and zeros are discussed in Appendix 4A at the back of this chapter.

4.2.1 Loop Transfer Function $G_L(s)$

It is the closed-loop response (with the feedback in place) that we need to optimize. Using linear control theory, we can achieve this objective by ensuring certain characteristics of the loop transfer function $G_L(s)$. In the control block diagram of Figure 4.3, the loop transfer function (from point A to point B) is

$$G_L(s) = G_C(s)G_{PWM}(s)G_{PS}(s)k_{FB}. \tag{4.2}$$

4.2.2 Crossover Frequency f_c of $G_L(s)$

In order to define a few necessary control terms, we will consider a generic Bode plot of the loop transfer function $G_L(s)$ in terms of its magnitude and phase angle, shown in Figure 4.4 as a function of frequency. The frequency at which the gain equals unity (that is $|G_L(s)| = 0\,\mathrm{dB}$) is defined as the crossover frequency f_c (or ω_c). This crossover frequency is a good indicator of the bandwidth of the closed-loop feedback system, which determines the speed of the dynamic response of the control system to various disturbances.

4.2.3 Phase and Gain Margins

For the closed-loop feedback system to be stable, at the crossover frequency f_c, the phase delay introduced by the loop transfer function must be less than 180°. At f_c, the

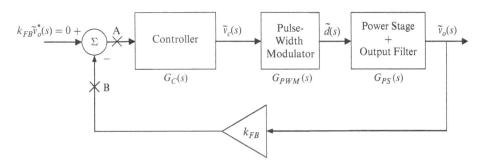

FIGURE 4.3 Small-signal control system representation.

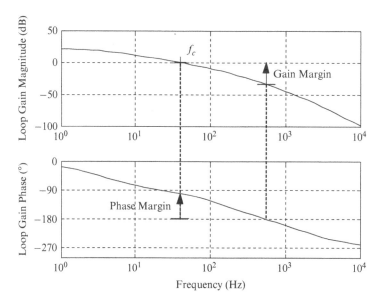

FIGURE 4.4   Definitions of crossover frequency, gain margin, and phase margin.

phase angle $\angle G_L(s)|_{f_c}$ of the loop transfer function $G_L(s)$, measured with respect to $-180°$, is defined as the phase margin (ϕ_{PM}) as shown in Figure 4.4:

$$\phi_{PM} = \angle G_L(s)|_{f_c} - (-180°) = \angle G_L(s)|_{f_c} + 180°. \tag{4.3}$$

Note that $\angle G_L(s)|_{f_c}$ is negative, but the phase margin in Equation (4.3) must be positive. Generally, feedback controllers are designed to yield a phase margin of approximately 60°, since much smaller values result in high overshoots and long settling times (oscillatory response) and much larger values in a sluggish response.

The gain margin is also defined in Figure 4.4, which shows that the gain margin is the value of the magnitude of the loop transfer function, measured below 0 dB, at the frequency at which the phase angle of the loop transfer function may (not always) cross $-180°$. If the phase angle crosses $-180°$, the gain margin should generally be in excess of 10 dB in order to keep the system response from becoming oscillatory due to parameter changes and other variations.

4.3   LINEARIZATION OF VARIOUS TRANSFER FUNCTION BLOCKS

To be able to apply linear control theory in the feedback controller design, it is necessary that all the blocks in Figure 4.2 be linearized around their DC steady-state operating point, as shown by transfer functions in Figure 4.3.

4.3.1   Linearizing the Pulse-Width Modulator

In the feedback control, a high-speed PWM integrated circuit such as the UC3824 [1] from Unitrode/Texas Instruments may be used. Functionally, within this PWM IC shown in Figure 4.5a, the control voltage $v_c(t)$ generated by the error amplifier is compared with a ramp signal v_r with a constant amplitude $\hat{V}_r$ at a constant switching frequency f_s, as shown in Figure 4.5b. The output switching signal is represented by the switching function $q(t)$, which equals 1 if $v_c(t) \geq v_r$ and is 0 otherwise. The switch duty ratio in Figure 4.5b is given as

$$d(t) = \frac{v_c(t)}{\hat{V}_r}. \tag{4.4}$$

In terms of a disturbance around the DC steady-state operating point, the control voltage can be expressed as

$$v_c(t) = V_c + \tilde{v}_c(t). \tag{4.5}$$

Substituting Equation (4.5) into Equation (4.4),

$$d(t) = \underbrace{\frac{V_c}{\hat{V}_r}}_{D} + \underbrace{\frac{\tilde{v}_c(t)}{\hat{V}_r}}_{\tilde{d}(t)}. \tag{4.6}$$

In Equation (4.6), the second term on the right side equals $\tilde{d}(t)$, from which the transfer function of the PWM IC is

$$G_{PWM}(s) = \frac{\tilde{d}(s)}{\tilde{v}_c(s)} = \frac{1}{\hat{V}_r}. \tag{4.7}$$

It is a constant gain transfer function, as shown in Figure 4.5c in the Laplace domain.

Example 4.1
In PWM ICs, there is usually a DC voltage offset in the ramp voltage, and instead of $\hat{V}_r$ as shown in Figure 4.5b, a typical valley-to-peak value of the ramp signal is defined. In the PWM IC UC3824, this valley-to-peak value is 1.8 V. Calculate the linearized transfer function associated with this PWM-IC.

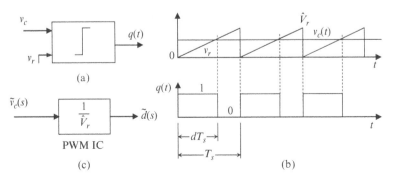

FIGURE 4.5   PWM waveforms.

Solution   The DC offset in the ramp signal does not change its small-signal transfer function. Hence, the peak-to-valley voltage can be treated as $\hat{V}_r$. Using Equation (4.7),

$$G_{PWM}(s) = \frac{1}{\hat{V}_r} = \frac{1}{1.8} = 0.556. \tag{4.8}$$

4.3.2   Linearizing the Power Stage of DC-DC Converters in CCM

To design feedback controllers, the power stage of the converters must be linearized around the steady-state DC operating point, assuming a small-signal disturbance. Figure 4.6a shows the average model of the switching power-pole, where the subscript "*vp*" refers to the voltage port and "*cp*" to the current port. Each average quantity in Figure 4.6a can be expressed as the sum of its steady-state DC value (represented by an uppercase letter) and a small-signal perturbation (represented by a tilde "~" on top):

$$\begin{aligned}
d(t) &= D + \tilde{d}(t) \\
\bar{v}_{vp}(t) &= V_{vp} + \tilde{v}_{vp}(t) \\
\bar{v}_{cp}(t) &= V_{cp} + \tilde{v}_{cp}(t) \\
\bar{i}_{vp}(t) &= I_{vp} + \tilde{i}_{vp}(t) \\
\bar{i}_{cp}(t) &= I_{cp} + \tilde{i}_{cp}(t).
\end{aligned} \tag{4.9}$$

Utilizing the voltage and current relationships between the two ports in Figure 4.6a and expressing each variable as in Equation (4.9),

$$V_{cp} + \tilde{v}_{cp} = (D + \tilde{d})(V_{vp} + \tilde{v}_{vp}), \tag{4.10a}$$

and

$$I_{vp} + \tilde{i}_{vp} = (D + \tilde{d})(I_{cp} + \tilde{i}_{cp}). \tag{4.10b}$$

Equating the perturbation terms on both sides of the above equations,

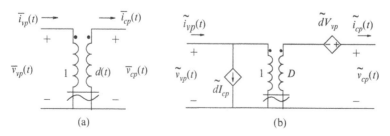

(a)   (b)

FIGURE 4.6   Linearizing the switching power-pole.

$$\tilde{v}_{cp}(t) = D\tilde{v}_{vp} + V_{vp}\tilde{d} + \tilde{d}\,\tilde{v}_{vp}, \tag{4.11a}$$

$$\tilde{i}_{vp}(t) = D\tilde{i}_{cp} + I_{cp}\tilde{d} + \tilde{d}\,\tilde{i}_{cp}. \tag{4.11b}$$

The two equations above are linearized by neglecting the products of small-perturbation terms. The resulting linear equations are

$$\tilde{v}_{cp}(t) = D\tilde{v}_{vp} + V_{vp}\tilde{d}, \tag{4.12}$$

and

$$\tilde{i}_{vp} = D\tilde{i}_{cp} + I_{cp}\tilde{d}. \tag{4.13}$$

Equations (4.12) and (4.13) can be represented by means of an ideal transformer shown in Figure 4.6b, which is a linear representation of the power pole for small signals around a steady-state operating point given by D, V_{vp}, and I_{cp}.

The average representations of buck, boost, and buck-boost converters are shown in Figure 4.7a. Replacing the power pole in each of these converters by its small-signals linearized representation, the resulting circuits are shown in Figure 4.7b, where the perturbation $\tilde{v}_{in}$ is zero-based on the assumption of a constant DC input voltage V_{in}, and the output capacitor ESR is represented by r. Note that in boost

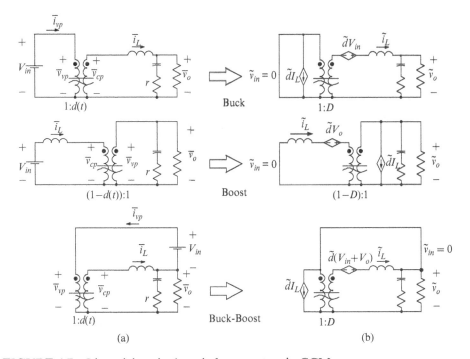

FIGURE 4.7 Linearizing single-switch converters in CCM.

converters, since the transistor is in the bottom position of the switching power-pole, d in Figure 4.6a needs to be replaced by $(1-d)$. Substituting d with $(D+\tilde{d})$ results in $(1-d)=(1-D)-\tilde{d}$. Therefore, D in Equations (4.12) and (4.13) needs to be replaced by $(1-D)$ and $\tilde{d}$ by $(-\tilde{d})$.

As fully explained in Appendix 4B, all three circuits for small-signal perturbations in Figure 4.7b have the same form as shown in Figure 4.8. In this equivalent circuit, the effective inductance L_e is the same as the actual inductance L in the buck converter, since in both states of a buck converter in CCM, L and C are always connected together. However, in boost and buck-boost converters, these two elements are not always connected, resulting in L_e to be $L/(1-D)^2$ in Figure 4.8:

$$L_e = L \quad \text{(Buck)}; \quad L_e = \frac{L}{(1-D)^2} \quad \text{(Boost and Buck - Boost)}. \qquad (4.14)$$

Transfer functions of the three converters in CCM from Appendix 4B are repeated below:

$$\frac{\tilde{v}_o}{\tilde{d}} = \frac{V_{in}}{LC} \frac{1+srC}{s^2 + s\left(\frac{1}{RC}+\frac{r}{L}\right)+\frac{1}{LC}} \qquad \text{(Buck)}, \qquad (4.15)$$

$$\frac{\tilde{v}_o}{\tilde{d}} = \frac{V_{in}}{(1-D)^2}\left(1-s\frac{L_e}{R}\right) \frac{1+srC}{L_eC\left[s^2 + s\left(\frac{1}{RC}+\frac{r}{L_e}\right)+\frac{1}{L_eC}\right]} \qquad \text{(Boost)}, \qquad (4.16)$$

$$\frac{\tilde{v}_o}{\tilde{d}} = \frac{V_{in}}{(1-D)^2}\left(1-s\frac{DL_e}{R}\right) \frac{1+srC}{L_eC\left[s^2 + s\left(\frac{1}{RC}+\frac{r}{L_e}\right)+\frac{1}{L_eC}\right]} \qquad \text{(Buck - Boost)}. \qquad (4.17)$$

In the above power-stage transfer functions in CCM, there are several characteristics worth noting. There are two poles created by the low-pass L-C filter in Figure 4.8, and the capacitor ESR r results in a zero. In boost and buck-boost converters, their transfer functions depend on the steady-state operating value D. They also have a right-half-plane zero, whose presence can be explained by the fact that in these converters, increasing the duty ratio for increasing the output, for example, initially has an opposite consequence by isolating the input stage from the output load for a longer time.

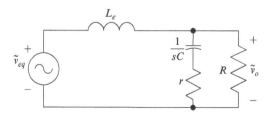

FIGURE 4.8   Small-signal equivalent circuit for buck, boost, and buck-boost converters.

4.3.2.1   Using Computer Simulation to Obtain $\tilde{v}_o / \tilde{d}$

Transfer functions given by Equations (4.15) through (4.17) provide theoretical insight into the converter operation. However, the Bode plots of the transfer function can be obtained with similar accuracy by means of linearization and AC analysis, using a computer program such as LTspice. The converter circuit is simulated as shown in Figure 4.9 in the example below for a frequency-domain AC analysis, using the switching power-pole average model discussed in Chapter 3 and shown in Figure 4.6a. The duty cycle perturbation $\hat{d}$ is represented as an AC source whose frequency is swept over several decades of interest and whose amplitude is kept constant, for example, at 1 V. In such a simulation, LTspice first calculates voltages and currents at the DC steady-state operating point, linearizes the circuit around this DC bias point, and then performs the AC analysis.

Example 4.2

A buck converter has the following parameters and is operating in CCM: $L = 100\,\mu\mathrm{H}$, $C = 697\,\mu\mathrm{F}$, $r = 0.1\,\Omega$, $\tilde{d}$, $V_{in} = 30\,\mathrm{V}$, and $P_o = 36\,\mathrm{W}$. The duty ratio D is adjusted to regulate the output voltage $V_o = 12\,\mathrm{V}$. Obtain the gain and the phase of the power stage $G_{PS}(s)$ for frequencies ranging from 1 Hz to 100 kHz.

Solution   The LTspice circuit is shown in Figure 4.9 where the DC voltage source $\{D\}$, representing the duty ratio D, establishes the DC operating point. The duty ratio perturbation $\tilde{d}$ is represented as an AC source whose frequency is swept over several decades of interest, keeping the amplitude constant. (Since the circuit is linearized before the AC analysis, the best choice for the AC source amplitude is 1 V.) The switching power-pole is represented by an ideal transformer, which consists of two dependent sources: a dependent current source and a dependent voltage source. The circuit parameters are specified by means of parameter blocks within LTspice.

The Bode plot of the frequency response is shown in Figure 4.10. It shows that at the crossover frequency $f_c = 1\mathrm{kHz}$ selected in the next example, Example 4.3, the power stage has $|G_{PS}(s)|_{f_c} = 24.66$ dB and $\angle G_{PS}(s)|_{f_c} = -138°$. We will make use of these values in Example 4.3.

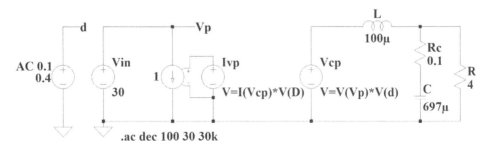

FIGURE 4.9   LTspice circuit model for a buck converter.

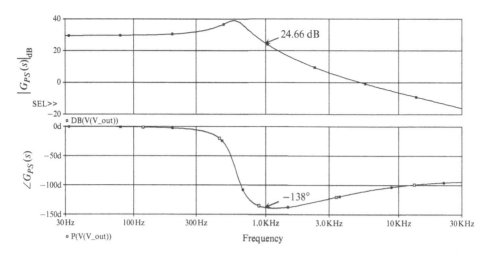

FIGURE 4.10　The gain and the phase of the power stage $G_{PS}(s)$.

4.4　FEEDBACK CONTROLLER DESIGN IN VOLTAGE-MODE CONTROL

The feedback controller design is presented by means of a numerical example to regulate the buck converter described earlier in Example 4.2. The controller is designed for the continuous conduction mode (CCM) at full load, which, although not optimum, is stable in DCM.

Example 4.3
Design the feedback controller for the buck converter described in Example 4.2. The PWM IC is as described in Example 4.1. The output voltage-sensing network in the feedback path has a gain $k_{FB} = 0.2$. The steady-state error is required to be zero, and the phase margin of the loop transfer function should be 60° at as high a crossover frequency as possible.

Solution　In deciding on the transfer function $G_C(s)$ of the controller, the control objectives translate into the following simultaneous characteristics of the loop transfer function $G_L(s)$, from which $G_C(s)$ can be designed:

1. The crossover frequency f_c of the loop gain is as high as possible to result in a fast response of the closed-loop system.
2. The phase angle of the loop transfer function has the specified phase margin, typically 60° at the crossover frequency, so that the response in the closed-loop system settles quickly without oscillations.
3. The phase angle of the loop transfer function should not drop below −180° at frequencies below the crossover frequency.

The Bode plot for the power stage is obtained earlier, as shown in Figure 4.10 of Example 4.2. In this Bode plot, the phase angle drops toward −180° due to the two poles of the L-C filter shown in the equivalent circuit of Figure 4.8 and confirmed by the transfer function of Equation (4.15). Beyond the L-C filter resonance frequency, the phase angle increases toward −90° because of the zero introduced by the output capacitor ESR in the transfer function of the power stage. We should not rely on this capacitor ESR, which is not accurately known and can have a large variability.

A simple procedure based on the K-factor approach [2] is presented below, which lends itself to a straightforward step-by-step design. For the reasons given below, the transfer function $G_C(s)$ of the controller is selected to be of the form in Equation (4.18), and its Bode plot is shown in Figure 4.11.

$$G_C(s) = \frac{k_c}{s} \frac{\left(1 + s/\omega_z\right)^2}{\left(1 + s/\omega_p\right)^2}.$$
(4.18)

To yield a zero steady-state error, $G_C(s)$ contains a pole at the origin, which introduces a −90° phase shift in the loop transfer function. The phase of the transfer function peaks at the geometric mean $\sqrt{f_z f_p}$ of the zero and pole frequencies, as shown in Figure 4.11, where f_z and f_p are chosen such that their geometric mean $\sqrt{f_z f_p}$ is equal to the loop crossover frequency f_c.

The crossover frequency f_c of the loop is chosen beyond the L-C resonance frequency of the power stage, where, unfortunately, $\angle G_{PS}|_{f_c}$ has a large negative value. The sum of −90° (due to the pole at the origin in $G_C(s)$) and $\angle G_{PS}|_{f_c}$ is more negative than −180°. Therefore, to obtain a phase margin of 60° requires boosting the phase at f_c, by more than 90°, by placing two coincident zeroes at f_z to nullify the effect of the two poles in the power-stage transfer function $G_{PS}(s)$. Two coincident poles are placed at f_p ($> f_z$) to roll off the gain rapidly much before the switching frequency. The controller gain $|G_C(s)|_{f_c}$ is such that the loop gain equals unity at the crossover frequency.

The input specifications in determining the parameters of the controller transfer function in Equation (4.18) are f_c, ϕ_{boost} as shown in Figure 4.11, and the controller gain $|G_C(s)|_{f_c}$. A step-by-step procedure for designing $G_C(s)$ is described below.

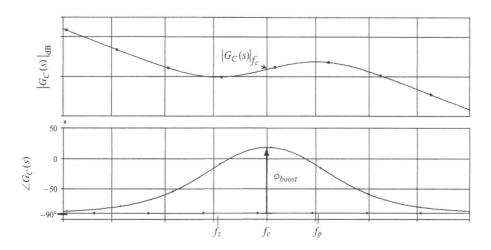

FIGURE 4.11 Bode plot of $G_C(s)$ in Equation (4.18).

Step 1: Choose the crossover frequency. Choose f_c to be *slightly* beyond the L-C resonance frequency $1/(2\pi\sqrt{LC})$, which in this example is approximately 600 Hz. Therefore, we will choose $f_c = 1\text{kHz}$. This ensures that the phase angle of the loop remains greater than $-180°$ at all frequencies below f_c.

Step 2: Calculate the needed phase boost. The desired phase margin is specified as $\phi_{PM} = 60°$. The required phase boost ϕ_{boost} at the crossover frequency is calculated as follows, noting that G_{PWM} and k_{FB} produce zero phase shift:

$$\angle G_L(s)|_{f_c} = \angle G_{PS}(s)|_{f_c} + \angle G_C(s)|_{f_c} \qquad \text{(from Equation 4.2)}, \qquad (4.19)$$

$$\angle G_L(s)|_{f_c} = -180° + \phi_{PM} \qquad \text{(from Equation 4.3)}, \qquad (4.20)$$

$$\angle G_C(s)|_{f_c} = -90° + \phi_{boost} \qquad \text{(from Figure 4.11)}. \qquad (4.21)$$

Substituting Equations (4.20) and (4.21) into Equation (4.19),

$$\phi_{boost} = -90° + \phi_{PM} - \angle G_{PS}(s)|_{f_c} . \qquad (4.22)$$

In Figure 4.10, $\angle G_{PS}(s)|_{fc} \simeq -138°$, substituting which in Equation (4.22), with $\phi_{PM} = 60°$, yields the required phase boost $\phi_{boost} = 108°$.

Step 3: Calculate the controller gain at the crossover frequency. From Equation (4.2) at the crossover frequency f_c,

$$\left|G_L(s)\right|_{f_c} = \left|G_C(s)\right|_{f_c} \times \left|G_{PWM}(s)\right|_{f_c} \times \left|G_{PS}(s)\right|_{f_c} \times k_{FB} = 1. \qquad (4.23)$$

In Figure 4.10, at $f_c = 1kHz$, $\left|G_{PS}(s)\right|_{f_c=1kHz} = 24.66\,dB = 17.1$. Therefore in Equation (4.23), using the gain of the PWM block calculated in Example 4.1,

$$\left|G_C(s)\right|_{f_c} \times \underbrace{0.556}_{|G_{PWM}(s)|_{f_c}} \times \underbrace{17.1}_{|G_{PS}(s)|_{f_c}} \times \underbrace{0.2}_{k_{FB}} = 1, \qquad (4.24)$$

or

$$\left|G_C(s)\right|_{f_c} = 0.5263. \qquad (4.25)$$

The controller in Equation (4.18) with two pole-zero pairs is analyzed in Appendix 4C. According to this analysis, the phase angle of $G_C(s)$ in Equation (4.18) reaches its maximum at the geometric mean frequency $\sqrt{f_z f_p}$, where the phase boost ϕ_{boost}, as shown in Figure 4.11, is measured with respect to $-90°$. By proper choice of the controller parameters, the geometric mean frequency is made equal to the crossover frequency f_c. We introduce a factor K_{boost} that indicates the geometric separation between poles and zeroes to yield the necessary phase boost:

$$K_{boost} = \sqrt{\frac{f_p}{f_z}}. \qquad (4.26)$$

As shown in Appendix 4C, K_{boost} can be derived in terms of ϕ_{boost} as follows:

$$K_{boost} = \tan\left(45° + \frac{\phi_{boost}}{4}\right).$$ (4.27)

Using the value of K_{boost} into Equation (4.26), and the fact that we will select $\sqrt{f_z f_p}$ to equal the chosen crossover frequency f_c, the pole and the zero frequencies in the controller can be calculated as follows:

$$f_z = \frac{f_c}{K_{boost}},$$ (4.28)

$$f_p = K_{boost}\, f_c.$$ (4.29)

From Equations (4.26), (4.28), and (4.29), the controller gain k_c in Equation (4.18) can be calculated at f_c as

$$k_c = \left|G_C(s)\right|_{f_c} \frac{\omega_z}{K_{boost}}.$$ (4.30)

Once the parameters in Equation (4.18) are determined, the controller transfer function can be synthesized by a single op-amp circuit shown in Figure 4.12. The choice of R_1 in Figure 4.12 is based on how much current can be drawn from the sensor output; other resistances and capacitances are chosen using the relationships derived in Appendix 4C and presented below:

$$\begin{aligned} &C_2 = \omega_z\,/(k_c\omega_p R_1), C_1 = C_2\left(\omega_p / \omega_z - 1\right), R_2 = 1/(\omega_z C_1), \\ &R_3 = R_1\,/(\omega_p / \omega_z - 1), C_3 = 1/(\omega_p R_3) \end{aligned}$$ (4.31)

In this numerical example with $f_c = 1\,\text{kHz}$, $\phi_{boost} = 108°$, and $\left|G_C(s)\right|_{f_c} = 0.5263$, we can calculate $K_{boost} = 3.078$ in Equation (4.27). Using Equations (4.28) through (4.30), $f_z = 324.9\,\text{Hz}$, $f_p = 3078\,\text{Hz}$, and $k_c = 349.1$. For the op-amp implementation, we will select $R_1 = 100\,\text{k}\Omega$. From Equation (4.31), $C_2 = 3.0\,\text{nF}$, $C_1 = 25.6\,\text{nF}$, $R_2 = 19.1\,\text{k}\Omega$, $R_3 = 11.8\,\text{k}\Omega$, and $C_3 = 4.4\,\text{nF}$.

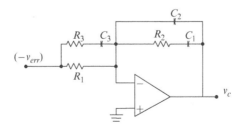

FIGURE 4.12 Controller implementation of $-G_c(s)$, using Equation (4.18), by an op-amp.

4.4.1 Simulation and Hardware Prototyping

The simulation of a voltage-mode control of buck converter using both LTspice and Workbench is demonstrated by means of an example:

Example 4.4

A buck converter is operating in CCM and has the following parameters: $L = 68\,\mu H$, $C = 490\,\mu F$, ESR $r = 0.28\,\Omega$, and load resistance $R = 8\,\Omega$. It is operating in DC steady state under the following conditions: $V_{in} = 15\,V$, $D = 0.5$, and $f_s = 100\,kHz$. For the switch and the diode, use the parameters given in the Appendix of Chapter 2. Design a voltage-mode controller to keep the voltage around this operating condition under varying input voltage and load. Assume that in the voltage feedback network, $k_{FB} = 1$. Simulate this converter using LTspice.

Solution The simulation file used in this example is available on the accompanying website. The controller parameters are computed using Workbench script in which Equation (4.19) through Equation (4.30) has been implemented as shown in Figure 4.13. Using a script file to auto-generate parameters helps quickly iterate through multiple design choices.

The crossover frequency is chosen at twice the *LC* resonance frequency, which comes out to be $f_c = 1.75\,kHz$. The phase margin is chosen to be $\phi_m = 60°$. The controller parameters computed by the script file are: $f_z = 950.1\,Hz$, $f_p = 3.19\,kHz$, and $k_c = 408.4$.

Using the above parameters, the controller is implemented by an op-amp in LTspice, as shown in Figure 4.14. The waveforms from the simulation of this model for a step-change in the load at 25 ms is shown in Figure 4.15.

```
Public Module Param ! Buck converter parameters
  ! Hardware components
  Public C as Native Double = 491µ ! output capacitance
  Public ESR as Native Double = 0.28 ! output capacitor ESR
  Public L as Native Double = 68µ ! inductance

  ! Operating condition
  Public Vin as Native Double = 15 ! input voltage
  Public Rl as Native Double = 8 ! load resistance
  Public Fs as Native Double = 100k ! switching frequency (Hz)
  Public D as Native Double = 0.5 ! duty cycle

  ! Controller parameters
  Private φm as Native Double = π / 3 ! phase margin (radians)
  ! cross-over frequency (Hz) choosen at 2x LC resonance frequency
  Private fc as Native Double = 2 * 1 / (2 * π * √(L * C))
  Private wc as Native Double = 2 * π * fc ! cross-over frequency (rad/s)

  ! Controller design
  ! converter magnitude and phase at fc
  Private GpsAtFc as Native Complex = Vin / L / C * (jwc * ESR * C + 1) /
  |                  ((jwc)² + jwc * (1 / Rl / C + ESR / L) + 1 / L / C)
  Private φboost as Native Double = -π / 2 + φm - ∠GpsAtFc
  Private Kboost as Native Double = Math:Tan(π / 4 + φboost / 4)

  ! Controller parameters
  Public wz as Native Double = wc / Kboost ! controller zero
  Public wp as Native Double = Kboost * wc ! controller pole
  Public kc as Native Double = 1 / ℳ(GpsAtFc) * wz / Kboost
End Module
```

FIGURE 4.13 Workbench script for computation of controller parameters.

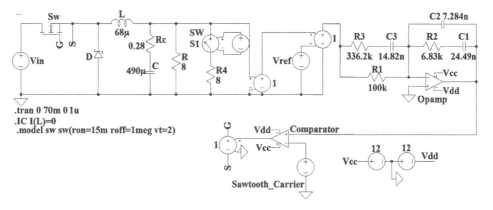

FIGURE 4.14   LTspice model.

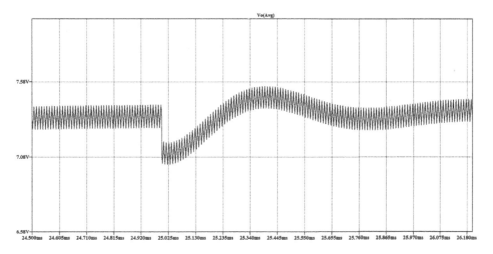

FIGURE 4.15   LTspice simulation results.

The same model can be implemented using Workbench, as shown in Figure 4.16. The advantage of using Workbench is that the controller can be implemented in the transfer function form as given by Equation (4.18) without having to convert it to an equivalent op-amp-based circuit. The implementation within the controller subsystem in Figure 4.16 is shown in Figure 4.17.

In the model, the output reference voltage is stepped from an initial value of 0 V to 7.5 V at time $t = 1\,\text{ms}$. The load is doubled, i.e. the load resistance is halved from $R = 8\,\Omega$ to $R = 4\,\Omega$ at $t = 5\,\text{ms}$. Finally, at $t = 8\,\text{ms}$ the input voltage is stepped up to $V_{in} = 22.5\,\text{V}$ from the previous value of $V_{in} = 15\,\text{V}$. Through this, the output voltage is maintained at $V_o = 7.5\,\text{V}$ by the controller as shown in Figure 4.18.

The Workbench model for implementing the above example in hardware using the Sciamble lab kit is shown in Figure 4.19. As mentioned earlier, the controller can be implemented directly in transfer function form, as shown in Figure 4.19. Unlike the op-amp-based controller implementation, where any changes to parameters would require changing physical components, the digital implementation shown here is

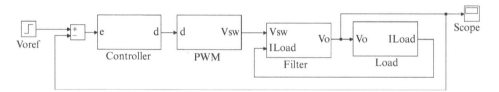

FIGURE 4.16   Workbench model.

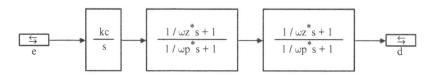

FIGURE 4.17   Controller subsystem.

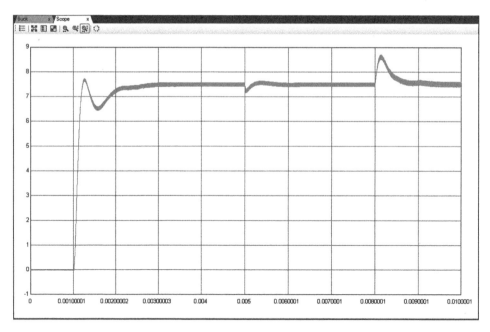

FIGURE 4.18   Output voltage waveform.

merely a matter of changing the numerical values in the software. This allows for rapid prototyping of various controllers using the same hardware.

The steady-state waveforms from running the buck converter using the Sciamble laboratory kit are shown in Figure 4.20. The converter output voltage settles down to the desired reference voltage as seen in Figure 4.20a and remains at the reference voltage for changing load resistance as seen in Figure 4.20b. Figure 4.20c shows the zoomed-in version of the waveforms over a few switching cycles. The step-by-step procedure for recreating the above hardware implementation is presented in [3].

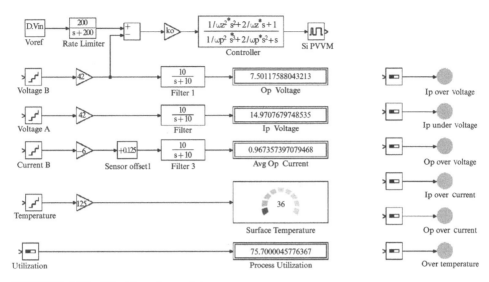

FIGURE 4.19   Workbench model.

4.5   PEAK-CURRENT MODE CONTROL

Current-mode control is often used in practice due to its many desirable features, such as simpler controller design and inherent current limiting. In such a control scheme, an inner control loop inside the outer voltage loop is used, as shown in Figure 4.21,

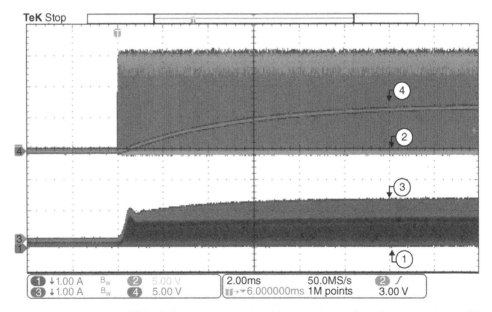

FIGURE 4.20   Workbench hardware results: (a) step change in reference voltage, (b) gradual change in load resistance, and (c) switching cycle waveforms. (1) input current, (2) switch-node voltage, (3) inductor current, and (4) output voltage. For clarity, see the waveforms in color in the Appendix on the accompanying website.

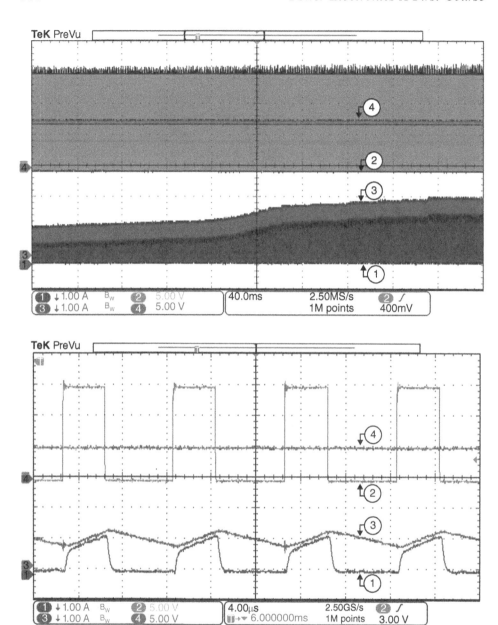

FIGURE 4.20   Continued

resulting in a peak-current-mode control system. In this control arrangement, another state variable, the inductor current, is utilized as a feedback signal.

The overall voltage-loop objectives in the current-mode control are the same as in the voltage-mode control discussed earlier. However, the voltage-loop controller here produces the reference value for the current that should flow through the inductor, hence the name current-mode control. There are two types of current-mode control:

1. Peak-current-mode control

2. Average-current-mode control.

In switch-mode DC power supplies, peak-current-mode control is invariably used, and therefore we will concentrate on it here. (We will examine the average-current-mode control in connection with the power-factor-correction circuits discussed in the next chapter.)

For the current loop, the outer voltage loop in Figure 4.21 produces the reference value i_L^* of the inductor current. This reference current signal is compared with the measured inductor current i_L to reset the flip-flop when i_L reaches i_L^*. As shown in Figures 4.21 and 4.22a, in generating i_L^*, the voltage controller output i_c is modified by a signal called the slope compensation, which is necessary to avoid oscillations at the sub-harmonic frequencies of f_s, particularly at the duty ratio $d > 0.5$. Generally, the slope of this compensation signal is less than one-half of the slope at which the inductor current falls when the transistor in the converter is turned off.

In Figure 4.22a, when the inductor current reaches the reference value, the transistor is turned off and is turned back on at a regular interval $T_s (= 1/f_s)$ set by the clock. For small perturbations, this current loop acts extremely fast, and it can be assumed ideal with a gain of unity in the small-signal block diagram of Figure 4.22b. The design of the outer voltage loop is described by means of the example below of a buck-boost converter operating in CCM.

Example 4.5

In this example, we will design a peak-current-mode controller for a buck-boost converter [4] that has the following parameters and operating conditions: $L = 100\,\mu H$, $C = 697\,\mu F$, $r = 0.01\,\Omega$, $f_s = 100\,kHz$, $V_{in} = 30\,V$. The output power $P_o = 18\,W$ in CCM and the duty ratio D is adjusted to regulate the output voltage $V_o = 12\,V$. The phase margin required for the voltage loop is $60°$. Assume that in the voltage feedback network, $k_{FB} = 1$.

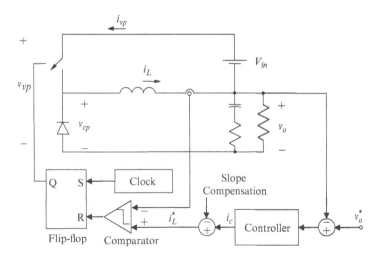

FIGURE 4.21 Peak current mode control.

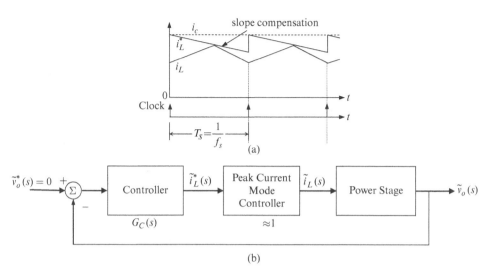

(b)

FIGURE 4.22   Peak-current-mode control with slope compensation.

Solution   In designing the outer voltage loop in Figure 4.22b, the transfer function needed for the power stage is $\tilde{v}_o / \tilde{i}_L$. This transfer function in CCM can be obtained theoretically. However, it is much easier to obtain the Bode plot of this transfer function by means of a computer simulation, similar to that used for obtaining the Bode plots of $\tilde{v}_o / \tilde{d}$ in Example 4.3 for a buck converter. The LTspice simulation diagram is shown in Figure 4.23 for the buck-boost converter, where, as discussed earlier, an ideal transformer is used for the average representation of the switching power-pole in CCM.

In Figure 4.23, the DC voltage source represents the switch duty ratio D and establishes the DC steady state, around which the circuit is linearized. In the AC

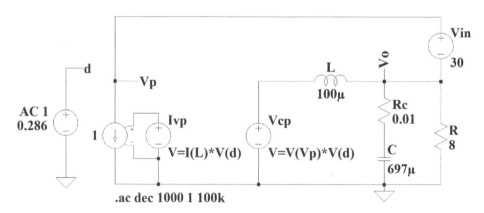

FIGURE 4.23   LTspice circuit for the buck-boost converter.

analysis, the frequency of the AC source, which represents the duty-ratio perturbation $\tilde{d}$, is swept over the desired range, and the ratio of $\tilde{v}_o(s)$ and $\tilde{i}_L(s)$ yields the Bode plot of the power stage $G_{PS}(s)$, as shown in Figure 4.24.

As shown in Figure 4.24, the phase angle of the power-stage transfer function levels off at approximately $-90°$ at $\simeq 1\,\text{kHz}$. The crossover frequency is chosen to be $5\,\text{kHz}$, at which in Figure 4.24, $\angle G_{PS}(s)|_{f_c} \simeq -90°$. The power-stage transfer function $\tilde{v}_o(s)/\tilde{i}_L(s)$ of buck-boost converters contains a right-half-plane zero in CCM, and the crossover frequency is chosen well below the frequency of the right-half-plane zero. To achieve the desired phase margin of $60°$, the controller transfer function is chosen as expressed below:

$$G_C(s) = \frac{k_c}{s} \frac{(1+s/\omega_z)}{(1+s/\omega_p)}. \tag{4.32}$$

To yield zero steady-state error, it contains a pole at the origin that introduces a $-90°$ phase angle. The phase-boost required from this pole-zero combination in Equation (4.32), using Equation (4.22) and $\angle G_{PS}(s)|_{f_c} \simeq -90°$, is $\phi_{boost} \simeq 60°$. Therefore, unlike the controller transfer function of Equation (4.18) for the voltage-mode control, only a single pole-zero pair is needed to provide a phase boost. In Equation (4.32), the zero and pole frequencies associated with the required phase boost can be derived, as shown in Appendix 4C, where K_{boost} is the same as in Equation (4.26), that is, $K_{boost} = \sqrt{f_p/f_z}$:

$$K_{boost} = \tan(45° + \frac{\phi_{boost}}{2}), \tag{4.33}$$

$$f_z = \frac{f_c}{K_{boost}}, \tag{4.34}$$

$$f_p = K_{boost} f_c \tag{4.35}$$

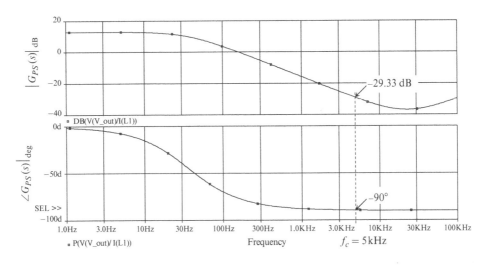

FIGURE 4.24 Bode plot of $\tilde{v}_o/\tilde{i}_L$.

$$k_c = \frac{\omega_z}{|G_{PS}(s)|_{f_c}}. \tag{4.36}$$

At the crossover frequency, as shown in Figure 4.24, the power stage transfer function has a gain $|G_{PS}(s)|_{f_c} = -29.33\,\text{dB}$. Therefore, at the crossover frequency, by definition, in Figure 4.22b,,

$$|G_C(s)|_{f_c} \times |G_{PS}(s)|_{f_c} = 1. \tag{4.37}$$

Hence,

$$|G_C(s)|_{f_c} = 29.33\,\text{dB} = 29.27. \tag{4.38}$$

Using the equations above for $f_c = 5\,\text{kHz}$ $\phi_{boost} \approx 60°$, and $|G_C(s)|_{f_c} = 29.27$, $K_{boost} = 3.732$ in Equation (4.33). Therefore, the parameters in the controller transfer function of Equation (4.32) are calculated as $f_z = 1340\,\text{Hz}$, $f_p = 18660\,\text{Hz}$, and $k_c = 246.4 \times 10^3$.

The transfer function of Equation (4.32) can be realized by an op-amp circuit shown in Figure 4.25. In the expressions derived in Appendix 4C, selecting $R_1 = 10\,\text{k}\Omega$ and using the transfer-function parameters calculated above, the component values in the circuit of Figure 4.25 are as follows:

$$C_2 = \frac{\omega_z}{\omega_p R_1 k_c} = 30\,p\text{F}$$
$$C_1 = C_2(\omega_p / \omega_z - 1) = 380\,p\text{F}. \tag{4.39}$$
$$R_2 = 1/(\omega_z C_1) = 315\,\text{k}\Omega$$

4.5.1   Simulation and Hardware Prototyping

The simulation of a peak-current-mode control of buck-boost converter using both LTspice and Workbench is demonstrated by means of an example:

Example 4.6

A buck-boost converter is operating in CCM and has the following parameters: $L = 68\,\mu\text{H}$, $C = 490\,\mu F$, ESR $r = 0.28\Omega$, and load resistance $R = 8\Omega$. It is operating in DC steady

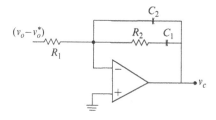

FIGURE 4.25   Controller implementation of $-G_c(s)$, using Equation (4.32), by an op-amp.

state under the following conditions: $V_{in} = 15\,\text{V}$, $V_o = 8$, and $f_s = 100\,\text{kHz}$. For the switch and the diode, use the parameters given in the Appendix of Chapter 2. Design a peak-current-mode controller to keep the voltage around this operating condition under varying input voltage and load. Assume that in the voltage feedback network, $k_{FB} = 1$. Simulate this converter using LTspice.

Solution   The simulation file used in this example is available on the accompanying website. The controller parameters are computed using Workbench script in which Equation (4.32) through Equation (4.38) has been implemented.

The crossover frequency is chosen to be $f_c = 1\,\text{kHz}$, below the frequency of the ESR zero, which occurs at $f_{z,esr} = r \times C = 1/(0.28 \times 490\mu \times 2\pi) = 1.16\,\text{kHz}$, to continue the gain roll-off at higher frequencies. The phase margin is chosen to be $\phi_m = 60°$. The controller parameters computed by the script file are: $f_z = 718.5\,\text{Hz}$, $f_p = 1.39\,\text{kHz}$, and $k_c = 11604.2$.

Using the above parameters, the controller is implemented by an op-amp in LTspice, as shown in Figure 4.26. The waveforms from the simulation of this model for a step-change in the load at 15 ms is shown in Figure 4.27.

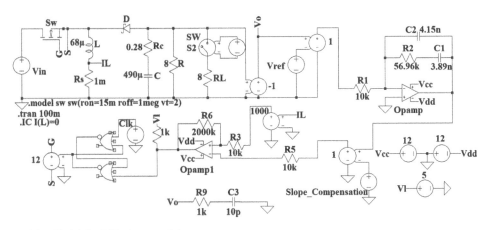

FIGURE 4.26   LTspice model.

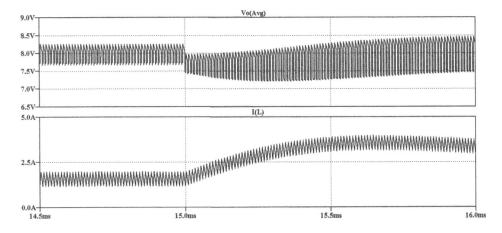

FIGURE 4.27   LTspice simulation results.

The same model can be implemented using Workbench, as shown in Figure 4.28. The voltage controller can be implemented in transfer function form, as given by Equation (4.32), as shown in Figure 4.29a, and the current controller as shown in Figure 4.29b.

In the model, the output reference voltage is stepped from an initial value of 0 V to 8 V at time $t = 1$ ms. The load is doubled, i.e. the load resistance is halved from $R = 8\,\Omega$ to $R = 4\,\Omega$ at $t = 5$ ms. Finally, at $t = 8$ ms the input voltage is stepped up to $V_{in} = 22.5\,$V from the previous value of $V_{in} = 15\,$V. Through this, the output voltage is maintained at $V_o = 7.5\,$V by the controller as shown in Figure 4.30.

The Workbench model for implementing the above example in hardware using the Sciamble lab kit is shown in Figure 4.19. As mentioned earlier, the controller can be implemented directly in transfer function form, as shown in Figure 4.31.

The steady-state waveforms from running the buck converter using the Sciamble laboratory kit are shown in Figure 4.32. The converter output voltage settles down to the desired reference voltage as seen in Figure 4.32a. Figure 4.32b shows the zoomed-in version of the waveforms over a few switching cycles. The step-by-step procedure for re-creating the above hardware implementation is presented in [4].

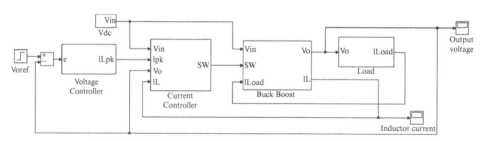

FIGURE 4.28   Workbench model.

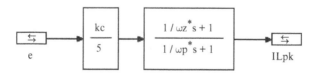

FIGURE 4.29A   Controller subsystem.

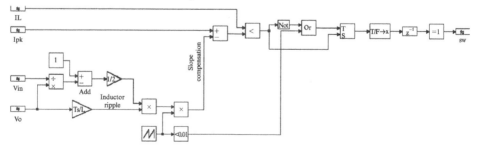

FIGURE 4.29B   Controller subsystem.

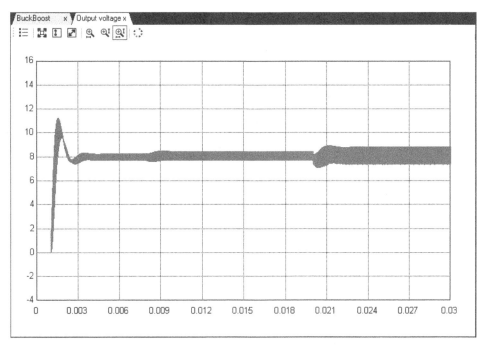

FIGURE 4.30 Output voltage waveform.

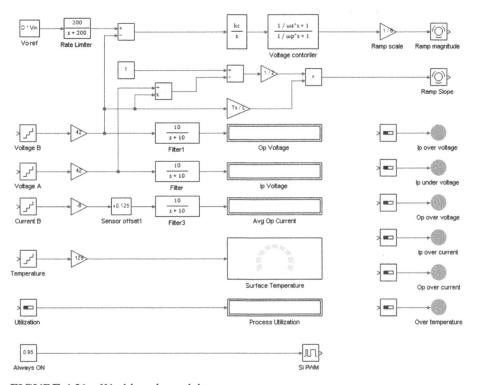

FIGURE 4.31 Workbench model.

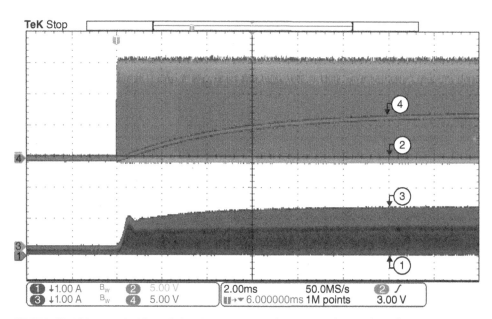

FIGURE 4.32A   Workbench hardware results for a step change in reference voltage:
(1) input current, (2) switch-node voltage, (3) inductor current, and (4) output voltage.
For clarity, see the waveforms in color in the Appendix on the accompanying website.

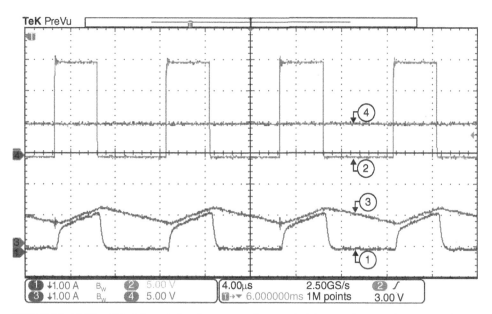

FIGURE 4.32B   Workbench hardware results—switching cycle waveforms: (1) input
current, (2) switch-node voltage, (3) inductor current, and (4) output voltage.

4.6   FEEDBACK CONTROLLER DESIGN IN DCM

In Sections 4.4 and 4.5, feedback controllers were designed for CCM operation of the converters. The procedure for designing controllers in DCM is the same, except that the average model of the power stage in LTspice simulations can be simply replaced by its model, which is also valid in DCM, as described in Chapter 3. This is illustrated for a buck-boost converter in the LTspice schematic of Figure 4.33, where the average model of the switching pole is valid for both CCM and DCM modes.

The Bode plot of the power stage $G_{PS}(s)$ in Figure 4.34 shows that in the DCM mode, as compared to the CCM mode, the phase plot appears as if one of the poles in the transfer function cancels out, making it easier to design the feedback controller in this mode.

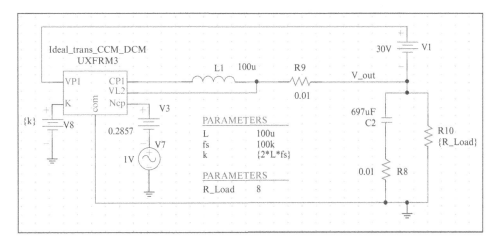

FIGURE 4.33   LTspice circuit for the buck-boost converter in both CCM and DCM modes.

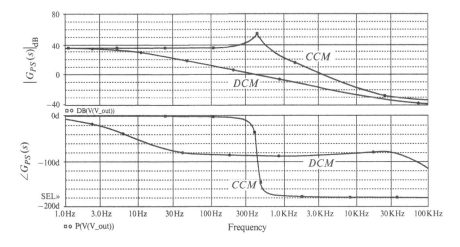

FIGURE 4.34   The gain and phase of the power stage $G_{PS}(s) = \tilde{v}_o / \tilde{d}(s)$ in CCM and DCM.

REFERENCES

1. PWM Controller ICs: Digital power control drivers & powertrain modules product selection | TI.com. https://www.ti.com/power-management/digital-power/digital-power-control-drivers-powertrain-modules/products.html.

2. H. Dean Venable, "The K-Factor: A New Mathematical Tool for Stability Analysis and Synthesis," Proceedings of Powercon 10. http://www.venable.biz.

3. Buck Converter Voltage-mode Control Lab Manual. https://sciamble.com/resources/pe-drives-lab/basic-pe/buck-voltage-mode-control.

4. Buck-Boost Converter Current-mode Control Lab Manual. https://sciamble.com/resources/pe-drives-lab/basic-pe/buck-boost-current-mode-control.

PROBLEMS

4.1   In Example 4.3, plot the gain and the phase of the open-loop transfer function $G_L(s)$.

4.2   In a voltage mode-controlled DC-DC converter the loop transfer function has the crossover frequency $f_c = 2\,\text{kHz}$. The power stage transfer function has a phase angle of $-160°$ at the crossover frequency. Calculate w_z and w_p in the voltage controller transfer function of Equation (4.18), if the required phase margin is $60°$.

4.3   In the above problem the power stage has a gain equal to 20 at the crossover frequency, $k_{FB} = 0.2$, and $G_{PWM} = 0.6$. Calculate k_c in the voltage controller transfer function of Equation (4.18).

4.4   In Example 4.4, plot the gain and the phase of the open-loop transfer function $G_L(s)$.

4.5   In a peak-current-mode-controlled DC-DC converter, the loop crossover frequency in the outer voltage loop is 10 kHz. At this crossover frequency, the power stage in Figure 4.22b has the gain of 0.1, and the phase angle of $-80°$. Calculate f_z, f_p and k_c in the controller transfer function of Equation (4.32) if the desired phase margin is $60°$.

4.6   Derive the transfer function $\tilde{v}_o(s)/\tilde{i}_L(s)$ for a buck converter in CCM.

4.7   Derive the transfer function $\tilde{v}_o(s)/\tilde{i}_L(s)$ for a boost converter in CCM.

4.8   Derive the transfer function $\tilde{v}_o(s)/\tilde{i}_L(s)$ for a buck-boost converter in CCM.

Simulation Problems

4.9   In a buck converter, various parameters are as follows: $L = 100\,\mu\text{H}$, $C = 697\,\mu\text{F}$ and $R_{Load} = 9.0\,\Omega$. The capacitor ESR is $0.1\,\Omega$. The input voltage $V_{in} = 24\,\text{V}$, the switching frequency $f_s = 100\,\text{kHz}$, and the output voltage $V_0 = 18\,\text{V}$.

(a)   Obtain the Bode plots for the transfer function $\dfrac{\tilde{v}_o}{\tilde{d}}(s)$ as shown in Figure 4.10 for the values given in this problem.

(b)   Obtain the gain and the phase of the transfer function $\dfrac{\tilde{v}_o}{\tilde{d}}(s)$ at the frequency of 1 kHz, which will be chosen as the crossover of the open-loop transfer function $G_L(s)$ in the next problem.

4.10   In the buck converter of Problem 4.9, design the feedback controller using the voltage mode, as shown in Figure 4.13, where $k_{fb} = 0.2$ and $G_{PWM}(s) = 0.556$. Choose the open-loop crossover frequency to be 1 kHz (or close to it) and the phase margin of 60°.

   (a)   This feedback controller is to be simulated using an op-amp, as shown in Figure 4.12. Obtain the output voltage response for a step change in load.

   (b)   Repeat part (a) with a phase margin of 45°. Compare the output voltage response with that of a phase margin of 60°.

   (c)   Repeat part (a) with a crossover frequency of 2 kHz and compare the response to that in part (a).

4.11   In a buck-boost converter, various parameters are as follows: $L = 100\,\mu H$, $C = 697\,\mu F$ and $R_{Load} = 200.0\,\Omega$. The capacitor ESR is 0.01ω. The input voltage $V_{in} = 24\,V$ the switching frequency $f_s = 100\,kHz$ and the switch duty ratio $D = 0.375$.

   (a)   Obtain the Bode plots for the transfer function $\dfrac{\tilde{v}_o}{\tilde{i}_L}(s)$ as shown in Figure 4.24 for the values given in this circuit.

   (b)   Obtain the gain and the phase of the transfer function $\dfrac{\tilde{v}_o}{\tilde{i}_L}(s)$ in part (a) at the frequency of 5 kHz, which will be chosen as the crossover of the open-loop transfer function $G_L(s)$ in the next problem.

4.12   In the buck-boost converter of Problem 4.11, design the feedback controller using the peak-current-mode, as shown in Figure 4.26. Choose the open-loop crossover frequency to be 5 kHz (or close to it) and the phase margin of 60°.

   (a)   This feedback controller is to be simulated using an op-amp, as shown in Figure 4.26. Obtain various parameters.

   (b)   Obtain the output voltage and the inductor current response for a step change in load.

4.13   In the buck-boost converter of Problem 4.11, obtain $\dfrac{\tilde{v}_o}{\tilde{d}}(s)$ under CCM and DCM modes of operation and compare the results.

APPENDIX 4A BODE PLOTS OF TRANSFER FUNCTIONS WITH POLES AND ZEROS

In this section, Bode plots of various transfer functions are presented as a review.

4A.1   A Pole in a Transfer Function

A transfer function with a pole at ω_p is expressed below

$$T(s) = \frac{1}{1 + s/\omega_p}, \tag{4A.1}$$

whose gain and phase plots in Figure 4A.1 show that the gain beyond the pole frequency of ω_p starts to change at a rate of −20 dB/decade and the phase angle falls to −90° approximately a decade later.

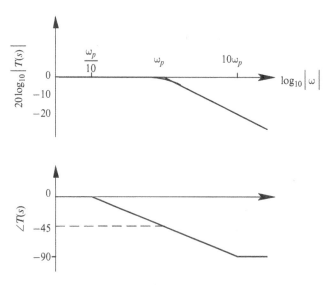

FIGURE 4A.1   Gain and phase plots of a pole.

4A.2   A Zero in a Transfer Function

The transfer function with a zero at a frequency of ω_z is expressed below:

$$T(s) = 1 + s/\omega_z, \tag{4A.2}$$

whose gain and phase plots in Figure 4A.2 show that the gain beyond the frequency of ω_z starts to rise at a rate of 20 dB/decade and the phase angle rises to $+90°$ approximately a decade later.

4A.3   A Right-Hand-Plane (RHP) Zero in a Transfer Function

In boost and buck-boost DC-DC converters, transfer functions contain a so-called right-hand plane (RHP) zero, with a transfer function expressed below:

$$T(s) = 1 - \frac{s}{\omega_z}, \tag{4A.3}$$

whose gain and phase plots in Figure 4A.3 show that the gain beyond the frequency of ω_z starts to rise at a rate of 20 dB/*decade* while the phase angle drops to $-90°$ approximately a decade later. This RHP zero presents special challenges in designing feedback controllers in boost and buck-boost converters, as is discussed in this chapter.

4A.4   A Double Pole in a Transfer Function

In DC-DC converter transfer functions, presence of L-C filters introduces a double pole, which can be expressed as below:

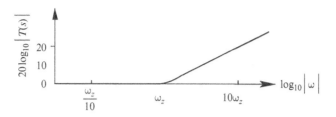

FIGURE 4A.2   Gain and phase plots of a zero.

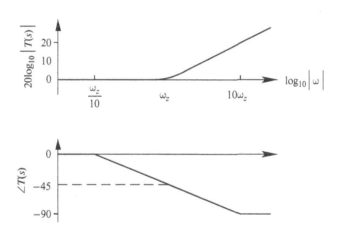

FIGURE 4A.3   Gain and phase plots of a right-hand side zero.

$$T(s) = \cfrac{1}{1 + \alpha s + \left(\dfrac{s}{\omega_o}\right)^2}, \tag{4A.4}$$

and whose gain and plots in Figure 4A.4 show that the gain beyond the frequency ω_o starts to fall at a rate of 40 dB$/$ *decade* and the phase angle falls toward $-180°$. These plots depend on the damping coefficient $\xi = (\alpha / 2)\omega_o$.

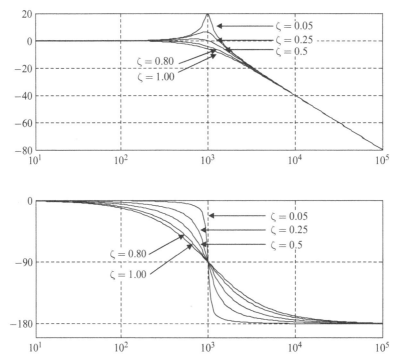

FIGURE 4A.4   Gain and phase plots of a double pole.

APPENDIX 4B TRANSFER FUNCTIONS IN CONTINUOUS CONDUCTION MODE (CCM)

In this section, we will derive the transfer function v_o/d for the three converters operating in CCM

4B.1   Buck Converters

From Figure 4.7, the small-signal diagram for a buck converter is shown in Figure 4B.1. The output stage impedance Z_{os} is defined as the parallel combination of the filter capacitor and the load resistance:

$$z_{os} = \frac{R(r + \frac{1}{sC})}{R + (r + \frac{1}{sC})} = R\frac{1 + srC}{1 + s(R + r)C}. \tag{4B.1}$$

In any practical converter, $r \ll R$, and therefore, $R + r \approx R$. Making use of this assumption in Equation (4B.1),

$$Z_{OS} \simeq R\frac{1 + srC}{1 + sRC}. \tag{4B.2}$$

Defining Z_{eff} as the sum of the filter inductor impedance sL and the output stage impedance Z_{os},

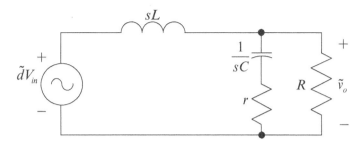

FIGURE 4B.1   Equivalent circuit of average buck converter.

$$\frac{Z_{OS}}{Z_{eff}} = \frac{1+srC}{LC\left[s^2 + s\left(\frac{1}{RC} + \frac{r}{L}\right) + \frac{1}{LC}\right]}. \tag{4B.3}$$

Therefore, in Figure 4B.1, by voltage division,

$$\frac{\tilde{v}_o}{\tilde{d}} = V_{in}\frac{Z_{os}}{z_{eff}} = V_{in}\frac{1+srC}{LC\left[s^2 + s\left(\frac{1}{RC} + \frac{r}{L}\right) + \frac{1}{LC}\right]}. \tag{4B.4}$$

4B.2   Boost Converter

From Figure 4.7, the small-signal diagram of a boost converter is shown in Figure 4B.2a. In this circuit, the DC steady-state operating point values can be calculated as follows:

$$I_o = \frac{V_o}{R}. \tag{4B.5}$$

Equating the input and the output power,

$$V_o I_o = V_{in} I_{in}. \tag{4B.6}$$

Substituting (Equation 4B.5) into (Equation 4B.6),

$$I_L = I_{in} = \frac{V_o I_o}{V_{in}} = \frac{V_o^2}{RV_{in}}. \tag{4B.7}$$

In Figure 4B.2a, the sub-circuit left of the marked terminals can be replaced by its Norton equivalent, as shown in Figure 4B.2b. The sub-circuit left of the transformer in Figure 4B.2b can be transformed to the right, as shown in Figure 4B.2c, where

$$L_e = \frac{L}{\left(1-D\right)^2}. \tag{4B.8}$$

The two current sources in Figure 4B.2c can be combined and using the Thevenin's equivalent, the equivalent voltage in Figure 4B.2d is

$$v_{eq} = \tilde{d}\,\frac{V_{in}}{(1-D)^2}\left(1 - \frac{sL_e}{R}\right) \qquad (4B.9)$$

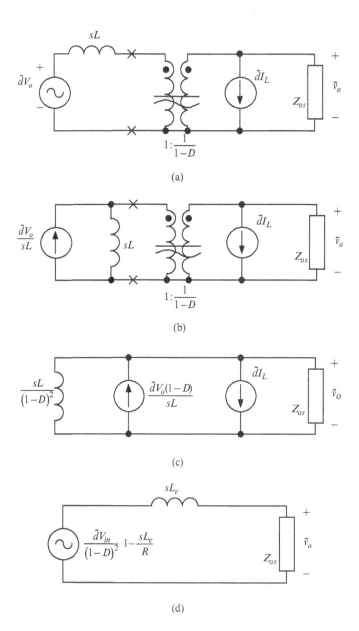

FIGURE 4B.2 Equivalent circuit of average boost converter.

Using the equivalent voltage in (Equation 4B.9) and applying the voltage division in the circuit of Figure 4B.2d,

$$\frac{\tilde{v}_o}{\tilde{d}} = \frac{V_{in}}{(1-D)^2}\left(1-\frac{sL_e}{R}\right)\frac{1+srC}{L_eC\left[s^2+s\left(\frac{1}{RC}+\frac{r}{L_e}\right)+\frac{1}{L_eC}\right]}. \tag{4B.10}$$

4B.3 Buck-Boost Converter

From Figure 4.7, the small-signal diagram of a buck-boost converter is shown in Figure 4B.3a. First, we will calculate the values of the needed quantities at the DC steady-state operating point.

In a buck-boost converter,

$$I_o = \frac{V_o}{R}, \tag{4B.11}$$

$$v_o = \frac{D}{1-D}V_{in}. \tag{4B.12}$$

Equating the input and the output power,

$$I_{in}V_{in} = V_oI_o \tag{4B.13}$$

and hence,

$$I_{in} = \frac{V_o^2}{RV_{in}}, \tag{4B.14}$$

$$I_L = \frac{V_{in}}{R}\frac{D}{(1-D)^2}. \tag{4B.15}$$

Considering the sub-circuit to the left of the marked terminals in Figure 4B.3a and drawn in Figure 4B.3b,

$$i_1 = i_2, \tag{4B.16}$$

where

$$i_1 = Di_2. \tag{4B.17}$$

(Equations 4B.16) and (4B.17) are valid in general only if $i_1 = i_2 = 0$. Therefore in Figure 4B.3b,

$$v_{oc} = \tilde{d}\frac{V_{in}}{(1-D)^2}. \tag{4B.18}$$

Shorting the terminals as shown in Figure 4B.3c,

$$i_{sc} = i_1 - Di_1 = (1-D)i_1. \tag{4B.19}$$

In Figure 4B.3c,

$$i_1 = \tilde{d}\frac{V_{in}}{(1-D)sL}. \tag{4B.20}$$

Substituting (Equation 4B.20) into (Equation 4B.19),

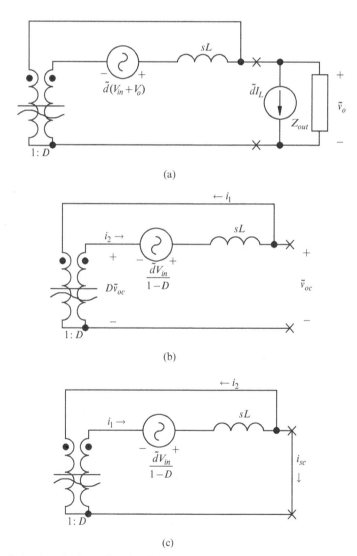

(a)

(b)

(c)

FIGURE 4B.3   Equivalent circuit of average buck-boost converter.

$$i_{sc} = \tilde{d}\frac{V_{in}}{sL}. \tag{4B.21}$$

From Figures 4B.3b and 4B.3c, and (Equations 4B.18) and (4B.21), the Thevenin impedance to the left of the marked terminals in Figure 4B.3a is

$$Z_{Th} = \frac{v_{oc}}{i_{sc}} = sL_e, \tag{4B.22}$$

where

$$L_e = \frac{L}{(1-D)^2}, \tag{4B.23}$$

$$v_{Th} = \tilde{d}\frac{V_{in}}{(1-D)^2}. \tag{4B.24}$$

With this Thevenin equivalent, the circuit of Figure 4B.3a, can be drawn as shown in Figure 4B.4a.

The sub-circuit to the left of the marked terminals can be represented by its Norton equivalent, as shown in Figure 4B.4b.

Combining the current sources and representing the sub-circuit in Figure 4B.4b by its Thevenin equivalent as shown in Figure 4B.4c,

$$v_{eq} = \tilde{d}\frac{V_{in}}{(1-D)^2}\left(1 - sD\frac{L_e}{R}\right). \tag{4B.25}$$

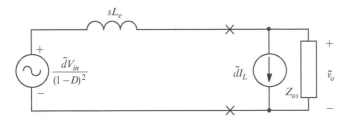

FIGURE 4B.4A Equivalent circuit of average buck-boost converter (contd.)

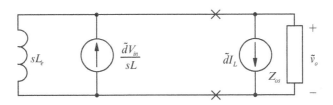

FIGURE 4B.4B Equivalent circuit of average buck-boost converter (contd.)

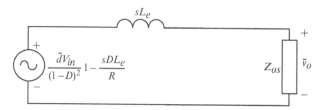

FIGURE 4B.4C Equivalent circuit of average buck-boost converter (contd.)

Hence,

$$\frac{\hat{v}_o}{d} = \frac{V_{in}}{(1-D)^2}\left(1 - sD\frac{L_e}{R}\right)\frac{1+srC}{L_eC\left[s^2 + s\left(\frac{1}{RC} + \frac{r}{L_e}\right) + \frac{1}{L_ec}\right]}. \qquad (4B.26)$$

APPENDIX 4C DERIVATION OF PARAMETERS OF THE CONTROLLER TRANSFER FUNCTIONS

4C.1 Controller Transfer Function with One Pole-Zero Pair

The controller transfer function given below consists of a pole at the origin and a pole-zero pair to provide phase boost:

$$G_C(s) = \frac{k_c}{s}\frac{1+s/\omega_z}{1+s/\omega_p}. \qquad (4C.1a)$$

To analyze this transfer function, the pole at the origin can be omitted since we know that it introduces a phase of $-90°$, by defining another transfer function as follows:

$$G'_c(s) = k_c\frac{1+s/\omega_z}{1+s/\omega_p}, \qquad (4C.1b)$$

where

$$\phi = \angle G'_c(s) = \tan^{-1}\frac{\omega}{\omega_p}. \qquad (4C.2)$$

4C.1.1 Frequency at which f_{boosl} Occurs

The maximum angle f_{boosl} provided by the controller occurs at the geometric mean of the zero and pole frequencies, as shown below. (This geometric mean frequency is made to coincide with $w = w_c$ where w_c is the cross-over frequency.) To find the frequency at which f_{boosl} occurs, we will set the derivative of the phase angle to zero.

Therefore,

$$\frac{d}{d\omega}\phi = \frac{1}{\omega_z}\frac{1}{\left(1+\dfrac{\omega^2}{\omega_z^2}\right)} - \frac{1}{\omega_p}\frac{1}{\left(1+\dfrac{\omega^2}{\omega_p^2}\right)} = 0, \qquad (4\text{C}.3)$$

or

$$\frac{\omega_z}{\left(\omega^2+\omega_z^2\right)} - \frac{\omega_p}{\left(\omega^2+\omega_p^2\right)} = 0, \qquad (4\text{C}.4)$$

$$\left(\omega^2 - \omega_z\omega_p\right)(\omega_z - \omega_p) = 0. \qquad (4\text{C}.5)$$

From (Equation 4C.5),

$$\omega = \sqrt{\omega_z\omega_p}, \qquad (4\text{C}.6)$$

which shows that the phase angle of the controller transfer function reaches its maximum at the geometric-mean frequency.

4C.1.2   Deriving the Zero and Pole Frequencies

Substituting (Equation 4C.6) into (Equation 4C.2),

$$\phi_{boost} = \tan^{-1}\frac{\sqrt{\omega_z\omega_p}}{\omega_z} - \tan^{-1}\frac{\sqrt{\omega_z\omega_p}}{\omega_p}, \qquad (4\text{C}.7)$$

or

$$\phi_{boost} = \tan^{-1}\sqrt{\frac{\omega_p}{\omega_z}} - \tan^{-1}\sqrt{\frac{\omega_z}{\omega_p}}. \qquad (4\text{C}.8)$$

Note that $\tan^{-1}x = \cot^{-1}\left(\dfrac{1}{x}\right)$ and $\tan^{-1}y + \cot^{-1}y = \dfrac{\pi}{2}$. Therefore, in (Equation 4C.8),

$$\phi_{boost} = \tan^{-1}\sqrt{\frac{\omega_p}{\omega_z}} - \left(\frac{\pi}{2} - \tan^{-1}\sqrt{\frac{\omega_p}{\omega_z}}\right) = 2\tan^{-1}\sqrt{\frac{\omega_p}{\omega_z}} - \frac{\pi}{2}. \qquad (4\text{C}.9)$$

We will define an intermediate variable, called the K-factor, as

$$k_{boost} = \sqrt{\frac{\omega_p}{\omega_z}}. \qquad (4\text{C}.10)$$

Solving (Equations 4C.9) and (4C.10),

$$k_{boost} = \tan\left(\frac{\phi_{boost}}{2} + \frac{\pi}{4}\right), \qquad (4\text{C}.11)$$

or

$$k_{boost} = \tan\left(\frac{\phi_{boost}}{2} + 45°\right).$$ (4C.12)

4C.1.3 Realizing the Controller Transfer Function with a Single Op-Amp

The controller transfer function in (Equation 4C.1) can be realized by a single op-amp circuit as shown below.

In Figure 4C.1, obtaining the input-output relationship and comparing it with the transfer function of (Equation 4C.1),

$$k_c = \frac{1}{R_1(C_1 + C_2)} \quad \omega_z = \frac{1}{C_1 R_2} \quad \omega_p = \frac{C_1 + C_2}{R_2 C_1 C_2}.$$ (4C.13)

From (Equation 4C.13), in terms of Rj

$$C_2 = \frac{\omega_Z}{\omega_P R_1 k_c} \quad C_1 = C_2(\omega_P / \omega_z - 1) \quad R_2 = 1 / (\omega_Z C_1).$$ (4C.14)

4C.2 Controller Transfer Function with Two Pole-Zero Pairs

The controller transfer function given below consists of a pole at the origin and two pole-zero pairs to provide phase boost

$$G_c'(s) = \frac{k_c}{s} \frac{(1 + s / \omega_z)^2}{(1 + s / \omega_p)^2}.$$ (4C.15)

To analyze this transfer function, the pole at the origin can be omitted since we know that it introduces a phase of −90°, by defining another transfer function as follows:

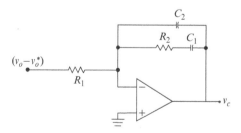

FIGURE 4C.1 Controller implementation of −G_c (s) , using Eq. 4-C1(a), by an op-amp.

$$G'_c(s) = k_c \frac{(1+s/\omega_z)^2}{(1+s/\omega_p)^2}, \tag{4C.16}$$

where

$$\phi = \angle G''_c(s) = 2\tan^{-1}\frac{\omega}{\omega_z} - 2\tan^{-1}\frac{\omega}{\omega_p}. \tag{4C.17}$$

A derivation similar to Section 4C.1 shows that the phase peaks at a frequency f_c that is the geometric mean of the pole and zero frequencies, similar to that in Section 4C.1:

$$\omega_c = \sqrt{\omega_z\omega_p}. \tag{4C.18}$$

Next, we will use the trigonometric identity that

$$\tan^{-1}x - \tan^{-1}y = \tan^{-1}\left(\frac{x-y}{1+xy}\right), \tag{4C.19}$$

and from Equations (4C.17) and (4C.18), at frequency ω_c, the phase boost is

$$\phi_{boost} = \angle G'_c(s) = 2\tan^{-1}\left(\frac{\dfrac{\omega_c}{\omega_z} - \dfrac{\omega_c}{\omega_p}}{1+\dfrac{\omega_c^2}{\omega_z\omega_p}}\right), \tag{4C.20}$$

$$\tan\left(\frac{\phi_{boost}}{2}\right) = \frac{\omega_c(\omega_p - \omega_z)}{\omega_c\omega_p + \omega_c^2}, \tag{4C.21}$$

$$\tan\left(\frac{\phi_{boost}}{2}\right) = \frac{\omega_c(\omega_p - \omega_z)}{\omega_z\omega_p + \omega_c^2}, \tag{4C.22}$$

and using Equations (4C.20) and (4C.21),

$$k_{boost} = \tan\left(45° + \frac{\phi_{boost}}{4}\right). \tag{4C.23}$$

The controller transfer function in (Equation 4C.15) can be realized by a single op-amp circuit as shown below.

In Figure 4C.2, obtaining the input-output relationship and comparing it with the transfer function of (Equation 4C.15) in terms of Rj,

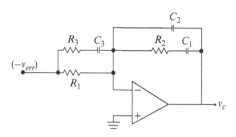

FIGURE 4C.2 Controller implementation of $-G_c(s)$, using Eq. 4-C15, by an op-amp.

$$c_2 = \omega_z / (k_c \omega_p R_1)$$
$$c_1 = c_2 (\omega_p / \omega_z - 1)$$
$$R_2 = 1 / (\omega_z c_1) \qquad\qquad (4C.24)$$
$$R_3 = R_1 / (\omega_p / \omega_z - 1)$$
$$C_3 = 1 / (\omega_p R_3).$$

5

RECTIFICATION OF UTILITY INPUT USING DIODE RECTIFIERS

As discussed in the introduction to Chapter 1, the role of power electronics is to facilitate power flow, often in a controlled manner, between two systems shown in Figure 5.1: one of them a "source" and the other a "load." Typically, power is provided by a single-phase or a three-phase utility source, for example, in adjustable-speed motor drives. (Of course, there are exceptions, for example, in wind turbines, where the wind-turbine generator is the source of power to the utility grid that acts like a "load.")

Such power-electronic interfaces often consist of a voltage-link structure, discussed in Section 1.5.1, where the input from the AC source is first rectified into a DC voltage across a large capacitor. If reversing power flow is not an objective, it is possible to rectify the AC input, single-phase or three-phase, by means of diode rectifiers discussed in this chapter. The knowledge of such systems is essential for learning about thyristor converters, discussed in Chapter 13, which are used in important applications such as high-voltage DC transmission (HVDC) systems.

5.1 INTRODUCTION

In diode rectifiers, unless corrective action is taken as described in the next chapter, power is drawn by means of highly distorted currents, which have a deleterious effect

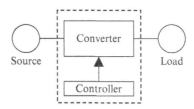

FIGURE 5.1 Block diagram of power electronic systems.

Power Electronics A First Course: Simulations and Laboratory Implementations, Second Edition.
Ned Mohan and Siddharth Raju.
© 2023 John Wiley & Sons, Inc. Published 2023 by John Wiley & Sons, Inc.
Companion Website: www.wiley.com/go/mohan/powerelectronics2e

on the power quality of the utility source. This issue and the basic principles of diode-rectifier operation are examined in this chapter.

5.2   DISTORTION AND POWER FACTOR

To quantify distortion in the current drawn by power electronic systems, it is necessary to define certain indices. As a base case, consider the linear $R-L$ load shown in Figure 5.2a, which is supplied by a sinusoidal source in steady state. The voltage and current phasors are shown in Figure 5.2b, where ϕ is the angle by which the current lags the voltage. Using RMS values for the voltage and current magnitudes, the average power supplied by the source is

$$P = V_s I_s \cos \phi \tag{5.1}$$

The power factor (PF) at which power is drawn is defined as the ratio of the real average power P to the product of the RMS voltage and the RMS current:

$$PF = \frac{P}{V_s I_s} = \cos \phi \quad \text{(using Equation 5.1)}, \tag{5.2}$$

where $V_s I_s$ is the apparent power. For a given voltage, from Equation (5.2), the RMS current drawn is

$$I_s = \frac{P}{V_s \cdot PF}. \tag{5.3}$$

This shows that the power factor PF and the current I_s are inversely proportional. This current flows through the utility distribution lines, transformers, and so on, causing losses in their resistances. This is the reason why utilities prefer unity power factor loads that draw power at the minimum value of the RMS current.

5.2.1   RMS Value of Distorted Current and the Total Harmonic Distortion (THD) [1]

The sinusoidal current drawn by the linear load in Figure 5.2 has zero distortion. However, power electronic systems with diode rectifiers as the front-end draw currents with a distorted waveform such as that shown by $i_s(t)$ in Figure 5.3a. The utility voltage $v_s(t)$ is assumed sinusoidal. The following analysis is general, applying to the utility supply that is either single-phase or three-phase, in which case the analysis is on a per-phase basis.

The current waveform $i_s(t)$ in Figure 5.3a repeats with a time period T_1. By Fourier analysis of this repetitive waveform, we can compute its fundamental frequency

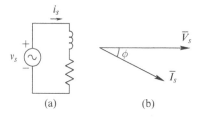

FIGURE 5.2   Voltage and current phasors in a simple *R-L* circuit.

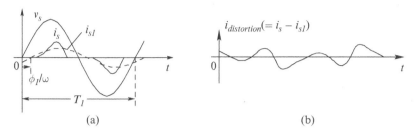

FIGURE 5.3 Current drawn by power electronics equipment with diode-bridge front-end.

$(=1/T_1)$ component $i_{s1}(t)$, shown dotted in Figure 5.3a. The distortion component $i_{distortion}(t)$ in the input current is the difference between $i_s(t)$ and the fundamental-frequency component $i_{s1}(t)$:

$$i_{distortion}(t) = i_s(t) - i_{s1}(t), \tag{5.4}$$

where $i_{distortion}(t)$ using Equation (5.4) is plotted in Figure 5.3b. This distortion component consists of components at frequencies that are the multiples of the fundamental frequency.

To obtain the RMS value of $i_s(t)$ in Figure 5.3a, we will apply the basic definition of RMS:

$$I_s = \sqrt{\frac{1}{T_1} \int_{T_1} i_s^2(t) \cdot dt}. \tag{5.5}$$

Using Equation (5.4),

$$i_s^2(t) = i_{s1}^2(t) + i_{distortion}^2(t) + 2i_{s1}(t) \times i_{distortion}(t). \tag{5.6}$$

In a repetitive waveform, the integral of the products of the two harmonic components (including the fundamental) at unequal frequencies, over the repetition time-period, equals zero:

$$\int_{T_1} g_{h_1}(t) \cdot g_{h_2}(t) \cdot dt = 0 \qquad h_1 \neq h_2. \tag{5.7}$$

Therefore, substituting Equation (5.6) into Equation (5.5) and making use of Equation (5.7) that implies that the integral of the third term on the right side of Equation (5.6) equals zero (assuming the average component to be zero),

$$I_s = \sqrt{\underbrace{\frac{1}{T_1} \int_{T_1} i_{s1}^2(t) \cdot dt}_{I_{s1}^2} + \underbrace{\frac{1}{T_1} \int_{T_1} i_{distortion}^2(t) \cdot dt}_{I_{distortion}^2} + 0}, \tag{5.8}$$

or

$$I_s = \sqrt{I_{s1}^2 + I_{distortion}^2}, \qquad (5.9)$$

where the RMS values of the fundamental-frequency component and the distortion component are as follows:

$$I_{s1} = \sqrt{\frac{1}{T_1} \int_{T_1} i_{s1}^2(t) \cdot dt}, \qquad (5.10)$$

and

$$I_{distortion} = \sqrt{\frac{1}{T_1} \int_{T_1} i_{distortion}^2(t) \cdot dt}. \qquad (5.11)$$

Based on the RMS values of the fundamental and the distortion components in the input current $i_s(t)$, a distortion index called the total harmonic distortion (THD) is defined in percentage as follows:

$$\% THD = 100 \times \frac{I_{distortion}}{I_{s1}}. \qquad (5.12)$$

Using Equation (5.9) into Equation (5.12),

$$\% THD = 100 \times \frac{\sqrt{I_s^2 - I_{s1}^2}}{I_{s1}}. \qquad (5.13)$$

The RMS value of the distortion component can be obtained based on the harmonic components (except the fundamental) as follows using Equation (5.7):

$$I_{distortion} = \sqrt{\sum_{h=2}^{\infty} I_{sh}^2}, \qquad (5.14)$$

where I_{sh} is the RMS value of the harmonic component "h."

5.2.1.1   Obtaining Harmonic Components by Fourier Analysis
By Fourier analysis, any distorted (non-sinusoidal) waveform $g(t)$ that is repetitive with a fundamental frequency f_1, for example, i_s in Figure 5.3a, can be expressed as a sum of sinusoidal components at the fundamental frequency and its multiples (harmonic frequencies):

$$g(t) = G_0 + \sum_{h=1}^{\infty} g_h(t) = G_0 + \sum_{h=1}^{\infty} \{a_h \cos(h\omega t) + b_h \sin(h\omega t)\}, \qquad (5.15)$$

where the average value G_0 is DC,

$$G_0 = \frac{1}{2\pi} \int_{2\pi}^{0} g(t) \cdot d(\omega t). \qquad (5.16)$$

The sinusoidal waveforms in Equation (5.15) at the fundamental frequency f_1 ($h = 1$) and the harmonic components at frequencies h times f_1 can be expressed as the sum of their cosine and sine components,

$$a_h = \frac{1}{\pi} \int_0^{2\pi} g(t)\cos(h\omega t)d(\omega t) \quad h = 1, 2, \ldots, \infty \tag{5.17}$$

$$b_h = \frac{1}{\pi} \int_0^{2\pi} g(t)\sin(h\omega t)d(\omega t) \quad h = 1, 2, \ldots, \infty. \tag{5.18}$$

The cosine and the sine components above, given by Equations (5.17) and (5.18), can be combined and written as a phasor in terms of its RMS value,

$$\bar{G}_h = G_h \angle \phi_h, \tag{5.19}$$

where the RMS magnitude in terms of the peak values a_h and b_h equals

$$G_h = \frac{\sqrt{a_h^2 + b_h^2}}{\sqrt{2}}, \tag{5.20}$$

and the phase ϕ_h can be expressed as

$$\tan \phi_h = \frac{-b_h}{a_h}. \tag{5.21}$$

It can be shown that the RMS value of the distorted function $g(t)$ can be expressed in terms of its average and the sinusoidal components as

$$G = \sqrt{G_0^2 + \sum_{h=1}^{\infty} G_h^2}. \tag{5.22}$$

In Fourier analysis, by appropriate selection of the time origin, it is often possible to make the sine or the cosine components in Equation (5.15) to be zero, thus considerably simplifying the analysis, as illustrated by a simple example below.

Example 5.1
A current i_s of a square waveform is shown in Figure 5.4a. Calculate and plot its fundamental frequency component and its distortion component. What is the %THD associated with this waveform?

Solution   From Fourier analysis, by choosing the time origin as shown in Figure 5.4a, $i_s(t)$ in Figure 5.4a can be expressed as

$$i_s = \frac{4}{\pi} I \left(\sin \omega_1 t + \frac{1}{3}\sin 3\omega_1 t + \frac{1}{5}\sin 5\omega_1 t + \frac{1}{7}\sin 7\omega_1 t + \ldots \right). \tag{5.23}$$

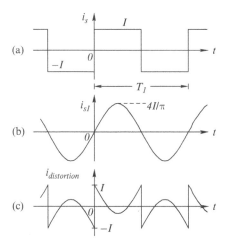

FIGURE 5.4 Example 5.1.

The fundamental frequency component and the distortion component are plotted in Figures 5.4b and 5.4c.

From Figure 5.4a, it is obvious that the RMS value I_s of the square waveform is equal to I. In the Fourier expression of Equation (5.23), the RMS value of the fundamental-frequency component is

$$I_{s1} = \frac{(4/\pi)}{\sqrt{2}} I = 0.9I$$

Therefore, the distortion component can be calculated from Equation (5.9) as

$$I_{distortion} = \sqrt{I_s^2 - I_{s1}^2} = \sqrt{I^2 - (0.9I)^2} = 0.436I.$$

Therefore, using the definition of THD,

$$\%\,THD = 100 \times \frac{I_{distortion}}{I_{s1}} = 100 \times \frac{0.436I}{0.9I} = 48.4\%.$$

5.2.2 The Displacement Power Factor (*DPF*) and Power Factor (*PF*)

Next, we will consider the power factor at which power is drawn by a load with a distorted current waveform such as that shown in Figure 5.3a. As before, it is reasonable to assume that the utility-supplied line-frequency voltage $v_s(t)$ is sinusoidal, with an RMS value of V_s and a frequency $f_1 \left(= \frac{\omega_1}{2\pi}\right)$. Based on Equation (5.7), which states that the product of the cross-frequency terms has a zero average, the average power P drawn by the load in Figure 5.3a is due only to the fundamental-frequency component of the current:

$$P = \frac{1}{T_1} \int_{T_1} v_s(t) \cdot i_s(t) \cdot dt = \frac{1}{T_1} \int_{T_1} v_s(t) \cdot i_{s1}(t) \cdot dt. \tag{5.24}$$

Therefore, in contrast to Equation (5.1) for a linear load, in a load that draws distorted current, similar to Equation (5.1),

$$P = V_s I_{s1} \cos \phi_1, \tag{5.25}$$

where ϕ_1 is the angle by which the fundamental-frequency current component $i_{s1}(t)$ lags behind the voltage, as shown in Figure 5.3a.

At this point, another term called the displacement power factor (*DPF*) needs to be introduced, where,

$$DPF = \cos \phi_1. \tag{5.26}$$

Therefore, using the *DPF* in Equation (5.25),

$$P = V_s I_{s1}(DPF). \tag{5.27}$$

In the presence of distortion in the current, the meaning and therefore the definition of the power factor, at which the real average power P is drawn, remains the same as in Equation (5.2), that is, the ratio of the real power to the product of the RMS voltage and the RMS current:

$$PF = \frac{P}{V_s I_s}. \tag{5.28}$$

Substituting Equation (5.27) for P into Equation (5.28),

$$PF = \left(\frac{I_{s1}}{I_s} \right)(DPF). \tag{5.29}$$

In linear loads that draw sinusoidal currents, the current-ratio (I_{s1} / I_s) in Equation (5.29) is unity, hence $PF = DPF$. Equation (5.29) shows the following: a high distortion in the current waveform leads to a low power factor, even if the *DPF* is high. Using Equation (5.13), the ratio (I_{s1}/I_s) in Equation (5.29) can be expressed in terms of the total harmonic distortion as

$$\frac{I_{s1}}{I_s} = \frac{1}{\sqrt{1 + \left(\frac{\% THD}{100} \right)^2}}. \tag{5.30}$$

Therefore, in Equation (5.29),

$$PF = \frac{1}{\sqrt{1 + \left(\frac{\% THD}{100} \right)^2}} \cdot DPF \tag{5.31}$$

The effect of THD on the power factor is shown in Figure 5.5 by plotting (PF / DPF) versus THD. It shows that even if the displacement power factor is unity, a total harmonic distortion of 100% (which is possible in power electronic systems

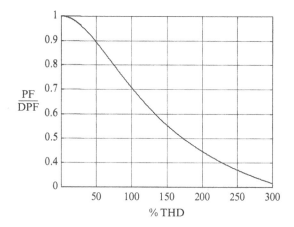

FIGURE 5.5   Relation between *PF/DPF* and THD.

unless corrective measures are taken) can reduce the power factor to approximately 0.7 (or $\frac{1}{\sqrt{2}} = 0.707$, to three decimal places, to be exact) which is unacceptably low.

5.2.3   Deleterious Effects of Harmonic Distortion and a Poor Power Factor

There are several deleterious effects of high distortion in the current waveform and the poor power factor that results due to it. These are as follows:

- Power loss in utility equipment such as distribution and transmission lines, transformers, and generators increases, possibly to the point of overloading them.
- Harmonic currents can overload the shunt capacitors used by utilities for voltage support and may cause resonance conditions between the capacitive reactance of these capacitors and the inductive reactance of the distribution and transmission lines.
- The utility voltage waveform will also become distorted, adversely affecting other linear loads, if a significant portion of the load supplied by the utility draws power by means of distorted currents.

5.2.3.1   Harmonic Guidelines

In order to prevent degradation in power quality, recommended guidelines (in the form of the IEEE-519) have been suggested by the IEEE (Institute of Electrical and Electronics Engineers). These guidelines place the responsibilities of maintaining power quality on the consumers and the utilities as follows: (1) on the power consumers, such as the users of power electronic systems, to limit the distortion in the current drawn, and (2) on the utilities to ensure that the voltage supply is sinusoidal with less than a specified amount of distortion.

The limits on current distortion placed by the IEEE-519 are shown in Table 5.1, where the limits on harmonic currents, as a ratio of the fundamental component, are specified for various harmonic frequencies. Also, the limits on the THD are specified. These limits are selected to prevent distortion in the voltage waveform of the utility supply. Therefore, the limits on distortion in Table 5.1 depend on the "stiffness" of the

TABLE 5.1 Harmonic current distortion (I_h / I_l)

I_{sc}/I_l	Odd Harmonic Order h (in %)					Total Harmonic Distortion (%)
	$h < 11$	$11 \leq h < 17$	$17 \leq h < 23$	$23 \leq h < 35$	$35 \leq h$	
<20	4.0	2.0	1.5	0.6	0.3	5.0
20–50	7.0	3.5	2.5	1.0	0.5	8.0
50–100	10.0	4.5	4.0	1.5	0.7	12.0
100–1000	12.0	5.5	5.0	2.0	1.0	15.0
>1000	15.0	7.0	6.0	2.5	1.4	20.0

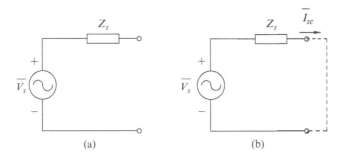

(a) (b)

FIGURE 5.6 (a) Utility supply; (b) short-circuit current.

utility supply, which is shown in Figure 5.6a by a voltage source $\bar{V}s$ in series with an internal impedance Z_s. An ideal voltage supply has zero internal impedance. In contrast, the voltage supply at the end of a long distribution line, for example, will have a large internal impedance.

To define the "stiffness" of the supply, the short-circuit current I_{sc} is calculated by hypothetically placing a short circuit at the supply terminals, as shown in Figure 5.6b. The stiffness of the supply must be calculated in relation to the load current. Therefore, the stiffness is defined by a ratio called the short-circuit ratio *(SCR)*:

$$\text{Short - Circuit - Ratio} \quad SCR = \frac{I_{sc}}{I_{s1}}, \tag{5.32}$$

where I_{s1} is the fundamental-frequency component of the load current. Table 5.1 shows that a smaller short-circuit ratio corresponds to lower limits on the allowed distortion in the current drawn. For the short-circuit ratio of less than 20, the total harmonic distortion in the current must be less than 5%. Power electronic systems that meet this limit would also meet the limits of more stiff supplies.

It should be noted that the IEEE-519 does not propose harmonic guidelines for individual pieces of equipment but rather for the aggregate of loads (such as in an industrial plant) seen from the service entrance, which is also the point of common coupling (PCC) with other customers. However, the IEEE-519 is frequently interpreted as the harmonic guidelines for specifying individual pieces of equipment such as motor drives. There are other harmonic standards, such as the IEC-1000, which apply to individual pieces of equipment.

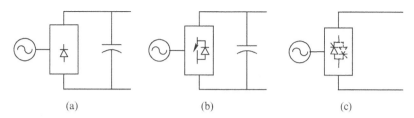

FIGURE 5.7   Front-end of power electronics equipment.

5.3   CLASSIFYING THE "FRONT-END" OF POWER ELECTRONIC SYSTEMS

Interaction between the utility supply and power electronic systems depends on the "front-ends" (within the power-processing units), which convert line-frequency AC into DC. These front-ends can be broadly classified as follows:

- Diode-bridge rectifiers (shown in Figure 5.7a) in which power flows only in one direction.
- Switch-mode converters (shown in Figure 5.7b) in which the power flow can reverse and the line currents are sinusoidal at the unity power factor.
- Thyristor converters (shown in Figure 5.7c) in which the power flow can be made bi-directional.

All of these front-ends can be designed to interface with single-phase or three-phase utility systems. In the following discussion, a brief description of the diode interface shown in Figure 5.7a is provided, supplemented by an analysis of results obtained through computer simulations. Interfaces using switch-mode converters in Figure 5.7b and thyristor converters in Figure 5.7c are discussed later in this book.

5.4   DIODE-RECTIFIER BRIDGE "FRONT-END"

Most power electronic systems use diode-bridge rectifiers, such as the one shown in Figure 5.7a, even though they draw currents with highly distorted waveforms and the power through them can flow only in one direction. In switch-mode DC power supplies, these diode-bridge rectifiers are supplemented by a power-factor-correction circuit to meet current harmonic limits, as discussed in the next chapter.

Diode rectifiers rectify line-frequency AC into DC across the DC-bus capacitor without any control over the DC-bus voltage. For analyzing the interaction between the utility and the power electronic systems, the switch-mode converter and the load can be represented by an equivalent resistance R_{eq} across the DC-bus capacitor. In our theoretical discussion, it is adequate to assume the diodes are ideal.

In the following subsections, we will consider single-phase as well as three-phase diode rectifiers operating in steady state, where waveforms repeat from one line-frequency cycle $T_1(=1/f_1)$ to the next.

5.4.1   Single-Phase Diode-Rectifier Bridge

At power levels below a few kW, for example, in residential applications, power electronic systems are supplied by a single-phase utility source. A commonly used

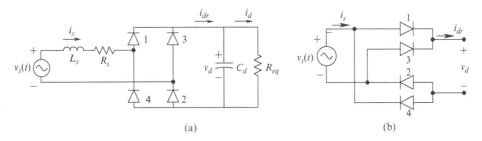

FIGURE 5.8 Full-bridge diode rectifier.

full-bridge rectifier circuit is shown in Figure 5.8a, in which L_s is the sum of the inductance internal to the utility supply and an external inductance, which may be intentionally added in series. Losses on the AC side can be represented by the series resistance R_s.

To understand the circuit operation, the rectifier circuit can be drawn as in Figure 5.8b, where the AC-side inductance and resistance have been ignored. The circuit consists of a top group and a bottom group of diodes. If the DC-side current i_{dr} is to flow, one diode from each group must conduct to facilitate the flow of this current. In the top group, both diodes have their cathodes connected together. Therefore, the diode connected to the most positive voltage will conduct; the other will be reverse biased. In the bottom group, both diodes have their anodes connected together. Therefore, the diode connected to the most negative voltage will conduct; the other will be reverse biased.

As an example, resistance R_d is connected across the terminals on the DC-side, as shown in Figure 5.9a. The circuit waveforms are shown Figure 5.9b. During the positive half-cycle of the input voltage v_s, diodes 1 and 2 conduct the DC-side current i_{dr}, equal to v_s / R_d, and the DC-side voltage is $v_d = v_s$. During the negative half-cycle of the input voltage v_s, diodes 3 and 4 conduct the DC-side current i_{dr}, equal to $|v_s|/R_d$, and the DC-side voltage is $v_d = |v_s|$.

The average value V_d of the voltage across the DC-side of the converter can be obtained by averaging the $v_d(t)$ waveform in Figure 5.9b over only one half-cycle (by symmetry):

$$V_d = \frac{1}{\pi}\int_0^\pi \hat{V}_s \sin\omega t \cdot d(\omega t) = \frac{2}{\pi}\hat{V}_s = 0.9V_s \,(\text{rms}), \qquad (5.33)$$

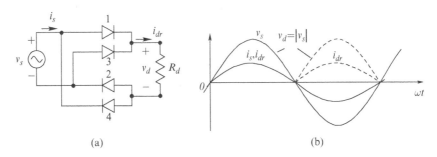

FIGURE 5.9 Full-bridge diode rectifier with resistive load.

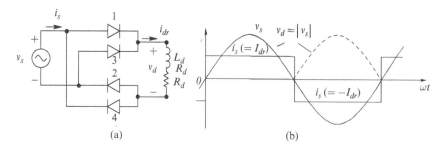

(a)   (b)

FIGURE 5.10   Full-bridge diode rectifier with an inductive load where $i_{dr}(t) \simeq I_{dr}$ (DC).

and

$$I_{dr} = \frac{V_d}{R_d} = \frac{0.9V_s(\text{rms})}{R_d} \tag{5.34}$$

As another example, consider the load on the DC-side to have a large inductance, as shown in Figure 5.10a, such that the DC-side current is essentially DC. Assuming that $i_{dr}(t) \simeq I_{dr}$ to be purely DC, the waveforms are as shown in Figure 5.10b.

Note that the waveform of $v_d(t)$ in Figure 5.10b is identical to that in Figure 5.9b for a resistive load. Therefore, the average value V_d of the voltage across the DC-side of the converter in Figure 5.10a is the same as in Equation (5.33), that is $V_d = 0.9V_s(\text{rms})$. Similarly, the average value of the DC-side current is related to V_d by R_d, as $I_{dr} = V_d / R_d$.

5.4.1.1   DC-Bus Capacitor to Achieve a Low-Ripple in the DC-Side Voltage

Figures 5.9b and 5.10b show that the DC-side voltage has a large ripple. To eliminate this, in order to achieve a voltage waveform that is fairly DC, a large filter capacitor is connected on the DC-side, as shown in Figure 5.8a. As shown in Figure 5.11, at the beginning of the positive half-cycle of the input voltage v_s, the capacitor is already charged to a DC voltage v_d. So long as v_d exceeds the input voltage magnitude, all diodes remain reverse biased, and the input current is zero. Power to the equivalent resistance R_{eq} is supplied by the energy stored in the capacitor up to t_1.

Beyond t_1, the input current $i_s(=i_{dr})$ increases, flowing through diodes D_1 and D_2. Beyond t_2, the input voltage becomes smaller than the capacitor voltage, and the input current begins to decline, falling to zero at t_3. Beyond t_3, until one half-cycle later than t_1, the input current remains zero, and the power to R_{eq} is supplied by the energy stored in the capacitor. At $(t_1 + T_1/2)$ during the negative half-cycle of the input

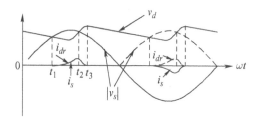

FIGURE 5.11   Waveforms for the full-bridge diode rectifier with a DC-bus capacitor.

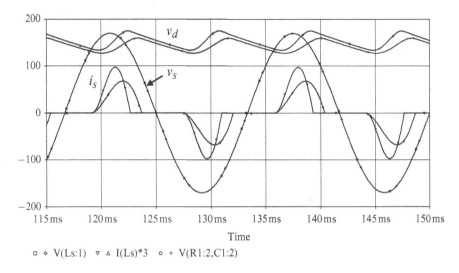

FIGURE 5.12   Single-phase diode-bridge rectification for two values of L_s.

voltage, the input current flows through diodes D_3 and D_4. The rectifier DC-side current i_{dr} continues to flow in the same direction as during the positive half-cycle; however, the input current $i_s = -i_{dr}$, as shown in Figure 5.11. Figure 5.12 shows waveforms obtained by LTspice simulations for two values of the AC-side inductance, with current THD of 86% and 62%, respectively (higher inductance reduces THD, as discussed in the next section).

The fact that i_{dr} flows in the same direction during both the positive and the negative half-cycles represents the rectification process. In the circuit of Figure 5.8a in steady state, all waveforms repeat from one cycle to the next. Therefore, the average value of the capacitor current over a line-frequency cycle must be zero so that the DC-bus voltage is in steady state. As a consequence, the average current through the equivalent load-resistance R_{eq} equals the average of the rectifier DC-side current; that is, $I_d = I_{dr}$.

5.4.1.2   Effects of L_s and C_d on the Waveforms and the THD

As Figures 5.11 and 5.12 show, power is drawn from the utility supply by means of a pulse of current every half-cycle. The larger the "base" of this pulse during which the current flows, the lower its peak value and the total harmonic distortion (THD). This pulse widening can be accomplished by increasing the AC-side inductance L_s. Another parameter under the designer's control is the value of the DC-bus capacitor C_d. At its minimum, it should be able to carry the ripple current in i_{dr} and in i_d (which in practice is the input DC-side current, with a pulsating waveform, of a switch-mode converter) and keep the peak-to-peak ripple in the DC-bus voltage to some acceptable value, for example, less than 5% of the DC-bus average value. Assuming that these constraints are met, the lower the value of C_d, the lower the THD in current and the higher the ripple in the DC-bus voltage.

In practice, it is almost impossible to meet the harmonic limits specified by IEEE-519 by using the above techniques. Rather, the power-factor-correction circuits described in the next chapter are needed to meet the harmonic specifications.

5.4.1.3 Simulation Using LTspice
The simulation of a single-phase diode-bridge rectifier is demonstrated by means of an example:

Example 5.2
In the single-phase diode-bridge rectifier shown in Figure 5.8a, $L_s = 1\,\text{mH}$, $R_s = 0.2\,\Omega$, $C_d = 1200\,\mu\text{F}$, and $R_{eq} = 15\,\Omega$. The supply voltage is $V_s = 120V$ RMS at 60 Hz. Simulate this rectifier using LTspice.

Solution The simulation file used in this example is available on the accompanying website. The LTspice model is shown in Figure 5.13, and the steady-state waveforms from the simulation of this model are shown in Figure 5.14.

5.4.2 Three-Phase Diode-Rectifier Bridge
It is preferable to use a three-phase utility source, except at a fractional kilowatt, if such a supply is available. A commonly used full-bridge rectifier circuit is shown in Figure 5.15a.

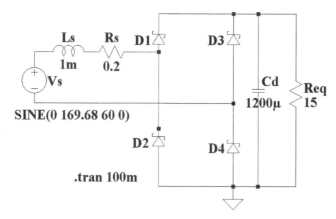

FIGURE 5.13 LTspice model.

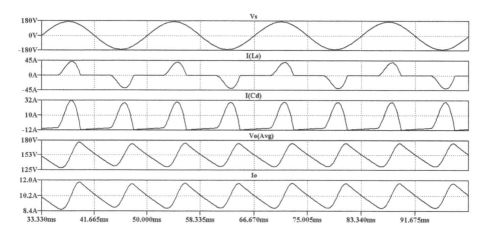

FIGURE 5.14 LTspice simulation results.

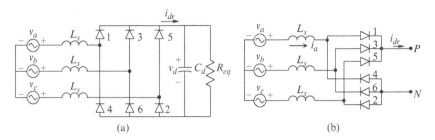

FIGURE 5.15 Three-phase diode bridge rectifier.

To understand the circuit operation, the rectifier circuit can be drawn as in Figure 5.15b. The circuit consists of a top group and a bottom group of diodes. Initially, L_s is ignored, and the DC-side current is assumed to flow continuously. At least one diode from each group must conduct to facilitate the flow of i_{dr}. In the top group, all diodes have their cathodes connected together. Therefore, the diode connected to the most positive voltage will conduct; the other two will be reverse biased. In the bottom group, all diodes have their anodes connected together. Therefore, the diode connected to the most negative voltage will conduct; the other two will be reverse biased.

Ignoring L_s and assuming that the DC-side current i_{dr} is a pure DC, the waveforms are as shown in Figure 5.16. In Figure 5.16a, the waveforms (identified by the dark portions of the curves) show that each diode, based on the principle described above, conducts during 120°. The diodes are numbered so that they begin conducting sequentially: 1, 2, 3, and so on. The waveforms for the voltages v_P and v_N, with respect to the source-neutral, consist of 120°-segments of the phase voltages, as shown in Figure 5.16a. The waveform of the DC-side voltage $v_d (= v_P - v_N)$ is shown in Figure 5.16b. It consists of 60°-segments of the line-line voltages supplied by the utility. Line currents on the AC-side are as shown in Figure 5.16c. For example, phase-a current flows for 120° during each half-cycle of the phase-a input voltage; it flows through diode D_1 during the positive half-cycle of v_a and through diode D_4 during the negative half-cycle.

The average value of the DC-side voltage can be obtained by considering only a 60°-segment in the 6-pulse (per line-frequency cycle) waveform shown in Figure 5.16b. Let us consider the instant of the peak in the 60°-segment to be the time-origin, with

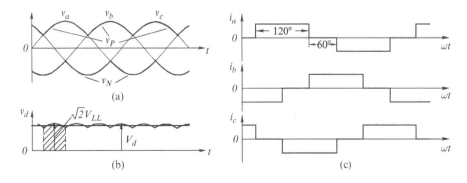

FIGURE 5.16 Waveforms in a three-phase rectifier (a constant i_{dr}).

$\widehat{V}_{LL}$ as the peak line-line voltage. The average value V_d can be obtained by calculating the integral from $\omega t = -\pi/6$ to $\omega t = \pi/6$ (the area shown by the hatched area in Figure 5.16b) and then dividing by the interval $\pi/3$:

$$V_d = \frac{1}{\pi/3} \int_{-\pi/6}^{\pi/6} \widehat{V}_{LL} \cos \omega t \cdot d(\omega t) = \frac{3}{\pi} \widehat{V}_{LL}. \qquad (5.35)$$

This average value is plotted as a straight line in Figure 5.16b.

5.4.2.1 Effect of DC-Bus Capacitor

In the three-phase rectifier of Figure 5.15a with the DC-bus capacitor filter, the input current waveforms obtained by computer simulations are shown in Figure 5.17.

Figure 5.17a shows that the input current waveform within each half-cycle consists of two distinct pulses when L_s is small. For example, in the i_a waveform during the positive half-cycle, the first pulse corresponds to the flow of DC-side current

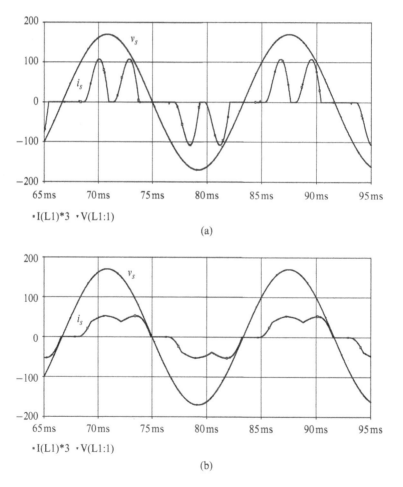

FIGURE 5.17 Effect of L_s variation (a) $L_s = 0.1\,\text{mH}$; (b) $L_s = 3\,\text{mH}$.

through the diode pair (D_1, D_6) and then through the diode pair (D_1, D_2). At larger values of L_s, within each half-cycle, the input current between the two pulses does not go to zero, as shown in Figure 5.17b.

The effects of L_s and C_d on the waveforms can be determined by a parametric analysis, similar to the case of single-phase rectifiers. The THD in the current waveform of Figure 5.17b is much smaller than in Figure 5.17a (23% versus 82%). The AC-side inductance L_s is required to provide a line-frequency reactance $X_{L_s} (= 2\pi f_1 L_s)$ that is typically greater than 2% of the base impedance Z_{base}, which is defined as follows:

$$\text{Base Impedance} \quad Z_{base} = 3\frac{V_s^2}{P}, \tag{5.36}$$

where P is the three-phase power rating of the power electronic system, and V_s is the RMS value of the phase voltage. Therefore, typically, the minimum AC-side inductance should be such that

$$X_{L_s} \geq (0.02 \times Z_{base}). \tag{5.37}$$

5.4.2.2 Simulation Using LTspice
The simulation of a three-phase diode-bridge rectifier is demonstrated by means of an example:

Example 5.3
In the three-phase diode-bridge rectifier shown in Figure 5.15a, $L_s = 0.1\text{mH}$, $C_d = 470\,\mu\text{F}$, and $R_{eq} = 15\,\Omega$. The supply voltage is $208V$ line-line RMS at $60\,\text{Hz}$. Simulate this rectifier using LTspice.

Solution The simulation file used in this example is available on the accompanying website. The LTspice model is shown in Figure 5.18, and the steady-state waveforms from the simulation of this model are shown in Figure 5.19.

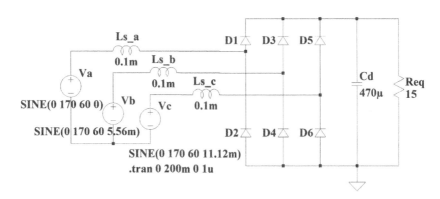

FIGURE 5.18 LTspice model.

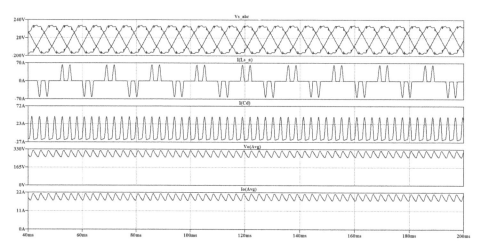

FIGURE 5.19   LTspice simulation results.

5.4.3   Comparison of Single-Phase and Three-Phase Rectifiers

Examination of single-phase and three-phase rectifier waveforms shows the differences in their characteristics. Three-phase rectification results in six identical "pulses" per cycle in the rectified DC-side voltage, whereas single-phase rectification results in two such pulses. Therefore, three-phase rectifiers are superior in terms of minimizing distortion in line currents and ripple across the DC-bus voltage. Consequently, as stated earlier, three-phase rectifiers should be used if a three-phase supply is available. However, three-phase rectifiers, just like single-phase rectifiers, are also unable to meet the harmonic limits specified by IEEE-519 unless corrective actions such as those described in Chapter 12 are taken.

5.5   MEANS TO AVOID TRANSIENT INRUSH CURRENTS AT STARTING

In power electronic systems with rectifier front-ends, it may be necessary to take steps to avoid a large inrush of current at the instant the system is connected to the utility source. In such power electronic systems, the DC-bus capacitor is very large and initially has no voltage across it. Therefore, at the instant the switch in Figure 5.20a is closed to connect the power electronic system to the utility source, a large current flows through the diode-bridge rectifier, charging the DC-bus capacitor.

This transient current inrush is highly undesirable; fortunately, several means of avoiding it are available. These include using a front-end that consists of thyristors, discussed in Chapter 14, or using a series semiconductor switch as shown in Figure 5.16b. At the instant of starting, the resistance across the switch lets the DC-bus capacitor to get charged without a large inrush current, and subsequently, the semiconductor switch is turned on to bypass the resistance. There are other techniques that can also be employed.

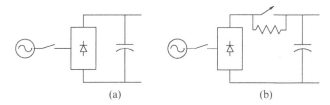

FIGURE 5.20 Means to avoid inrush current.

5.6 FRONT-ENDS WITH BI-DIRECTIONAL POWER FLOW

In stop-and-go applications such as elevators, it is cost effective to feed the energy recovered by regenerative braking of the motor drive back into the utility supply. Converter arrangements for such applications are considered in Chapter 12.

REFERENCES

1. N. Mohan, T.M. Undeland, and W.P. Robbins, *Power Electronics: Converters, Applications and Design*, 3rd Edition (New York: John Wiley & Sons, 2003).

PROBLEMS

5.1 In a single-phase diode rectifier bridge, $I_s = 10\,\mathrm{A\,(rms)}$, $DPF = 0.85$, and $DPF = 0.85$. Calculate $I_{distortion}$, $\%THD$, and PF.

5.2 In a single-phase diode bridge rectifier circuit, the following operating conditions are given: $V_s = 120V\,(\mathrm{rms})$, $P = 0.95\mathrm{kW}$, $I_{s1} = 10\,\mathrm{A}$, and $THD = 75\%$. Calculate the following: DPF, $I_{distortion}$, I_s, and PF.

5.3 In a single-phase rectifier the input current can be approximated by a triangular waveform every half cycle with a peak of $10\,\mathrm{A}$ and a base of $60°$. Calculate the RMS current through each diode.

5.4 In the above problem, calculate the ripple component in the DC-side current that will flow through the DC-side capacitor.

5.5 In a three-phase rectifier, if the input currents are of rectangular waveforms, as shown in Figure 5.16c, with an amplitude of $12\ A$, calculate I_s, I_{s1}, $I_{distortion}$, $\%THD$ and the power factor. Assume that $DPF = 1.0$, as associated with the current waveforms in Figure 5.14c.

5.6 In the above problem, calculate the three-phase power through the rectifier bridge if $V_{LL}\,(\mathrm{rms}) = 208\,\mathrm{V}$.

Simulation Problems

5.7 In the single-phase rectifier in Figure 5.8a, $L_s = 1\mathrm{mH}, R_s = 0.25\,\Omega, C_d = 1,000\,\mu F$ and $R_{eq} = 20.0\,\Omega$.
   (a) Obtain the %THD in the input current for the three values of the input inductance 1 mH, 3 mH, and 5 mH. Comment on the input current waveform as a function of the AC-side inductance in part (a).

(b) Measure the average and the peak-to-peak ripple in the output voltage for the three values of the input-side inductance in part (a), and comment on the output voltage waveform as a function of the AC-side inductance.

(c) Keeping the input inductance value as 1 mH, change the DC-side capacitance values to be $220\,\mu F$, $470\,\mu F$ and $1,000\,\mu F$. Measure the peak-to-peak ripple in the output capacitor voltage for these three values of the output capacitance.

5.8 In the three-phase rectifier in Figure 5.15a, $L_s = 0.1\,mH$, $C_d = 500\,\mu F$ and $R_{eq} = 16.5\,\Omega$.

(a) Obtain the %THD in the input current for the three values of the input inductance of 0.1 mH, 0.5 mH, and 1.0 mH.

(b) Comment on the input current waveform as a function of the AC-side inductance for values.

(c) Comment on the output voltage waveform as a function of the AC-side inductance. Measure the average and the peak-to-peak ripple in the output voltage for the three values of the input-side inductance.

(d) Measure the peak-to-peak ripple in the output capacitor current for the three values of the input-side inductance. What is the fundamental frequency of this current?

6

POWER-FACTOR-CORRECTION (PFC) CIRCUITS AND DESIGNING THE FEEDBACK CONTROLLER

In diode rectifiers discussed in Chapter 5, power is drawn by means of highly distorted currents, which have a deleterious effect on the power quality of the utility source. In single-phase diode-rectifier systems, a corrective action such as that described in this chapter is often taken. This discussion is also useful in the learning process since it shows a real-world application of DC-DC converters discussed in Chapter 3 and their control in Chapter 4.

6.1   INTRODUCTION

Technical solutions to the problem of distortion in the input current have been known for a long time. However, only recently has the concern about the deleterious effects of harmonics led to the formulation of guidelines and standards, which in turn have focused attention on ways of limiting current distortion.

In the following sections, power-factor-corrected (PFC) interface, as they are often called, are briefly examined for single-phase rectification, where it is assumed that the power needs to flow only in one direction, such as in DC power supplies. The three-phase front-ends in motor-drives applications may require bi-directional power flow capability. Such front-ends, which also allow unity power factor of operation, are discussed in Chapter 12.

6.2   OPERATING PRINCIPLE OF SINGLE-PHASE PFCS

The operating principle of a commonly used single-phase PFC is shown in Figure 6.1a, where, between the utility supply and the DC-bus capacitor, a boost DC-DC

Power Electronics A First Course: Simulations and Laboratory Implementations, Second Edition.
Ned Mohan and Siddharth Raju.
© 2023 John Wiley & Sons, Inc. Published 2023 by John Wiley & Sons, Inc.
Companion Website: www.wiley.com/go/mohan/powerelectronics2e

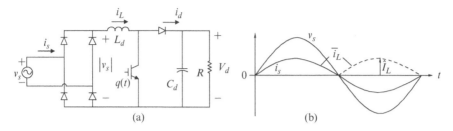

FIGURE 6.1 PFC circuit and waveforms.

converter is introduced. This boost converter consists of a MOSFET, a diode, and a small inductor L_d. By pulse-width-modulating the MOSFET at a constant switching frequency, the current i_L through the inductor L_d is shaped to have the full-wave-rectified waveform $\bar{i}_L(t) = \hat{I}_L |\sin \omega t|$, similar to $|v_s(t)|$, as shown in Figure 6.1b.

The inductor current contains a high switching-frequency ripple, which is removed by a small filter.

The input current i_s in the circuit of Figure 6.1a becomes sinusoidal and in phase with the supply voltage. In the boost converter, it is essential that the DC-bus voltage V_d be greater than the peak of the supply voltage $\hat{V}_s$:

$$V_d > \hat{V}_s. \tag{6.1}$$

Figure 6.2a shows the average model of the boost converter in the continuous-conduction mode (CCM) with $i_L > 0$ at all times. Neglecting a small voltage drop across the inductor and assuming the voltage across the capacitor to be a pure DC, in a boost converter, the voltage transfer ratio is

$$\frac{V_d}{|v_s|} = \frac{1}{1 - d(t)}, \tag{6.2}$$

and thus,

$$1 - d(t) = \frac{\hat{V}_s |\sin \omega t|}{V_d} \tag{6.3a}$$

and

$$d(t) = 1 - \frac{\hat{V}_s |\sin wt|}{V_d}. \tag{6.3b}$$

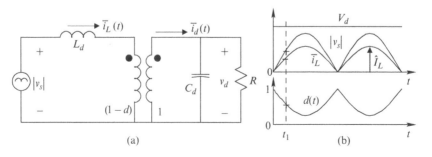

FIGURE 6.2 Average model in CCM $(i_L(t) > 0)$ and waveforms.

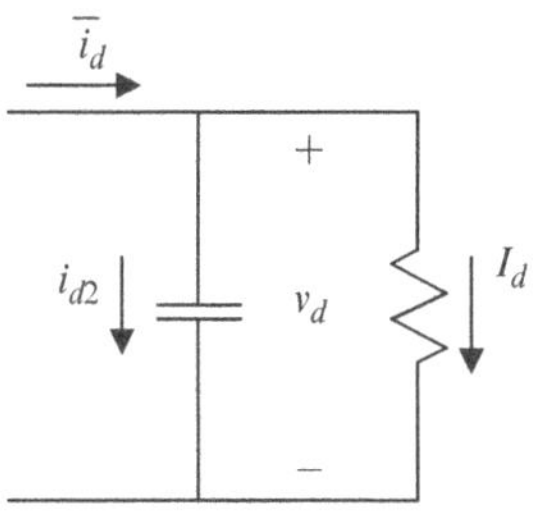

FIGURE 6.3 Current division in the output stage.

Equation (6.3) shows that $1 - d(t)$ of the switch varies sinusoidally during each half-cycle of the fundamental frequency, independent of the value of the inductor current, provided it is greater than zero. The switch duty ratio from Equation (6.3b) is plotted in Figure 6.2b.

The inductor current is shaped to be $\bar{i}_L(t) = \hat{I}_L |\sin \omega t|$. Therefore, the current $\bar{i}_d(t)$ can be calculated from the turns ratio of the ideal transformer in Figure 6.2a as

$$\bar{i}_d(t) = (1 - d)\bar{i}_L(t) = (1 - d)\hat{I}_L |\sin \omega t|. \tag{6.4a}$$

Substituting for $(1 - d)$ from Equation (6.3) in Equation (6.4),

$$\bar{i}_d(t) = \frac{\hat{V}_s}{V_d} \hat{I}_L |\sin \omega t|^2. \tag{6.4b}$$

Recognizing that in Equation (6.4), $|\sin \omega t|^2 = \sin^2 \omega t = \frac{1}{2} - \frac{1}{2}\cos 2\omega t$:

$$\bar{i}_d(t) = \underbrace{\frac{1}{2}\frac{\hat{V}_s}{V_d}\hat{I}_L}_{I_d} - \underbrace{\frac{1}{2}\frac{\hat{V}_s}{V_d}\hat{I}_L \cos 2\omega t}_{i_{d2}(t)}. \tag{6.5}$$

Equation (6.5) shows that the average current to the output stage consists of a DC component I_d and a component $i_{d2}(t)$ at the second-harmonic frequency.

Example 6.1
Derive $\bar{i}_d(t)$ in Equation (6.5) by equating input and output powers.

Solution   Assume that $v_s = \hat{V}_s \sin \omega t$ and $i_s = \hat{I}_s \sin \omega t$. Therefore, the input power $P_{in}(t) = v_s i_s = \hat{V}_s \hat{I}_s \sin^2 \omega t$. Recognizing that $\sin^2 \omega t = \frac{1}{2} - \frac{1}{2}\cos 2\omega t$, the input power is $P_{in}(t) = \hat{V}_s \hat{I}_s \sin^2 \omega t = \frac{1}{2}\hat{V}_s \hat{I}_s - \frac{1}{2}\hat{V}_s \hat{I}_s \cos 2\omega t$. The output power $p_o(t) = V_d \bar{i}_d$. Equating $p_{in}(t) = p_o(t)$, and recognizing that $\hat{I}_s = \hat{I}_L$

$$\bar{i}_d(t) = \underbrace{\frac{1}{2}\frac{\hat{V}_s \hat{I}_L}{V_d}}_{I_d} - \underbrace{\frac{1}{2}\frac{\hat{V}_s}{V_d}\hat{I}_L \cos 2\omega t}_{i_{d2}(t)}, \tag{6.6}$$

which is the same as Equation (6.5).

The output stage of the PFC is shown in Figure 6.3, where the DC component I_d of $\bar{i}_d$ flows through the load-equivalent resistor. In PFCs, the capacitor in the output stage is quite large, such that it is justifiable to approximate that the second-harmonic frequency component $i_{d2}(t)$ flows entirely through the output capacitor and none through the load-equivalent resistor. The peak value of $i_{d2}(t)$ is

$$\hat{I}_{d2} = \frac{1}{2}\frac{\hat{V}_s}{V_d}\hat{I}_L. \tag{6.7}$$

Based on the assumption that $i_{d2}(t)$ flows entirely through the output capacitor, the peak value of the second-harmonic frequency ripple in the output voltage can be calculated as follows, where the reactance of the capacitor at twice the line frequency is $1/(2\omega C)$:

$$\hat{V}_{d2} \simeq \left(\frac{1}{2\omega C}\right)\hat{I}_{d2} = \frac{\hat{I}_L}{4\omega C}\frac{\hat{V}_s}{V_d}. \tag{6.8}$$

$\hat{V}_{d2}$ depends inversely on the output capacitance C, and therefore, an appropriate value of C must be chosen to minimize this ripple.

Example 6.2
Calculate $\hat{V}_{d2}$ at full-load and the nominal input voltage for the parameters and operating values of a PFC given in Table 6.1 later on. Ignore the capacitor ESR.

Solution   Assuming the PFC to be lossless, $V_sI_s = P_o$. Therefore, using the values given in Table 6.1, $\hat{I}_L = \hat{I}_s = \sqrt{2}\frac{P_o}{V_s} = 2.946\,\text{A}$. $\hat{V}_s = \sqrt{2}\times120 = 169.7\,\text{V}$. Therefore, from Equation (6.8), the peak value of the second-harmonic frequency voltage is

$$\hat{V}_{d2} = \frac{\hat{I}_L}{4\omega C}\frac{\hat{V}_s}{V_d} \simeq 6\,\text{V}.$$

6.3   CONTROL OF PFCS

Figure 6.4 shows the PFC power circuit along with its control circuit in a block-diagram form. In controlling a PFC, the main objective is to draw a sinusoidal current in-phase with the utility voltage. The reference inductor current $i_L^*(t)$ is of the full-wave rectified form, similar to that in Figure 6.1b. The requirements on the form and the amplitude of the inductor current lead to two control loops, as shown in Figure 6.4, to pulse-width modulate the switch of the boost converter.

- The average inner current control loop ensures the form of $i_L^*(t)$ based on the template $\sin|\omega t|$ provided by measuring the rectifier output voltage $|v_s(t)|$.
- The outer voltage control loop determines the amplitude $\hat{I}_L$ of $i_L^*(t)$ based on the output voltage feedback. If the inductor current is insufficient for a given load supplied by the PFC, the output voltage will drop below its preselected reference value V_d^*. By measuring the output voltage and using it as the feedback signal, the voltage

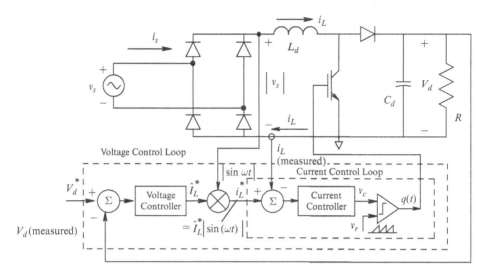

FIGURE 6.4 PFC control loops.

control loop adjusts the inductor current amplitude to bring the output voltage to its reference value. In addition to determining the inductor current amplitude, this voltage feedback control acts to regulate the output voltage of the PFC to the pre-selected DC voltage.

In Figure 6.4, the inner current-control loop is required to have a very high bandwidth compared to the outer voltage-control loop. Hence, each loop can be designed separately, similar to the approach taken in the peak-current-mode control discussed in Chapter 4.

6.4 DESIGNING THE INNER AVERAGE-CURRENT-CONTROL LOOP

The inner current control loop is shown within the inner dotted box in Figure 6.4. In order to follow the reference with as little THD as possible, an average-current-mode control is used with a high bandwidth, where the error between the reference $i_L^*(t)$ and the measured inductor current $i_L(t)$ is amplified by a current controller to produce the control voltage $v_c(t)$. This control voltage is compared with a ramp signal $v_r(t)$, with a peak of $\hat{V}_r$ at the switching frequency f_s in the PWM controller IC [1], to produce the switching signal $q(t)$.

Just the inner current control loop of Figure 6.4 can be simplified, as shown in Figure 6.5a. The reference input $i_L^*(t)$ varies with time, as shown in Figure 6.2b, where the corresponding $|v_s(t)|$ and $d(t)$ waveforms are also plotted. However, these quantities vary much more slowly compared to the current control-loop bandwidth, approximately 10 kHz in the numerical example considered later on. Therefore, at each instant of time, for example at t_1 in Figure 6.2b, the circuit of Figure 6.2a can be considered in a "DC" steady-state with the associated variables having values of $i_L(t_1)$, $|v_s(t_1)|$ and $d(t_1)$. This equilibrium condition varies slowly with time compared to the current-control-loop bandwidth, which is designed to be much larger. In the Laplace domain,

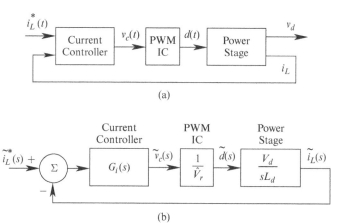

FIGURE 6.5   PFC current loop.

this current loop is shown in Figure 6.5b, as discussed below, where the tilde "~" on top represents small signal perturbations at very high frequencies in the range of the current-control-loop bandwidth, for example, 10 kHz.

6.4.1 $\tilde{d}(s)/\tilde{v}_c(s)$ for the PWM Controller

If $\hat{V}_r$ is the difference between the peak and the valley of the ramp voltage in the PWM-IC, then the small-signal transfer function of the PWM controller, as discussed in Chapter 4, is

$$\text{PWM Controller Transfer Function} \quad \frac{\tilde{d}(s)}{\tilde{v}_c(s)} = \frac{1}{\hat{V}_r}. \tag{6.9}$$

6.4.2 $\tilde{i}_L(s)/\tilde{d}(s)$ for the Boost Converter in the Power Stage

In spite of the varying DC steady-state operating point, the transfer function in the boost converter simplifies as follows at high frequencies at which the current-control-loop bandwidth is designed:

$$\frac{\tilde{i}_L(s)}{\tilde{d}(s)} \simeq \frac{V_d}{sL_d}. \tag{6.10}$$

This can be observed from the small-signal circuit for the boost converters in Figure 4.7b of Chapter 4, where at high frequencies, the capacitor acts as a short circuit, resulting in the above transfer function (ignoring capacitor ESR, and noting that V_o in Chapter 4 is V_d in this chapter). This conclusion can be confirmed by LTspice simulations that show that in the boost converter of a PFC, all the curves corresponding to various input voltages and the associated duty ratios merge at high frequencies to yield results similar to that of the transfer function in Equation (6.10).

6.4.3 Designing the Current Controller $G_{i(s)}$

The transfer function in Equation (6.10) is an approximation valid at high frequencies and not a pure integrator. Therefore, to have a high loop DC gain and a zero DC steady-state error in Figure 6.5b, the current controller transfer function $G_i(s)$ must have a pole at the origin. In the loop in Figure 6.5b, the phase due to the pole at origin in $G_i(s)$ and that of the power-stage transfer function (Equation 6.10) add up to $-180°$. Hence, $G_i(s)$, as in the peak-current mode control discussed in Chapter 4, includes a pole-zero pair that provides a phase boost, and hence the specified phase margin, for example 60° at the loop crossover frequency:

$$G_i(s) = \frac{k_c}{s} \frac{1 + s / \omega_z}{1 + s / \omega_p}, \tag{6.11}$$

where k_c is the controller gain. Knowing the phase boost, ϕ_{boost}, we can calculate the pole-zero locations to provide the necessary phase boost, as discussed in Chapter 4:

$$K_{boost} = \tan\left(45° + \frac{\phi_{boost}}{2}\right) \tag{6.12}$$

$$f_z = \frac{f_{ci}}{K_{boost}} \tag{6.13}$$

$$f_p = K_{boost} f_{ci}, \tag{6.14}$$

where f_{ci} is the crossover frequency of the current loop transfer function.

6.5 DESIGNING THE OUTER VOLTAGE-CONTROL LOOP

As mentioned earlier, the outer voltage-control loop is needed to determine the peak, $\hat{I}_L$, of the inductor current. In this voltage loop, the bandwidth is limited to approximately 15 Hz. The reason has to do with the fact that the output voltage across the capacitor contains a component v_{d2} as derived in Equation (6.8) at twice the line-frequency (at 120 Hz in 60-Hz line-frequency systems). This output voltage ripple must not be corrected by the voltage loop; otherwise, it will lead to a third-harmonic distortion in the input current, as explained in Appendix 6A at the end of this chapter.

In view of such a low bandwidth of the voltage-control loop (approximately three orders of magnitude below the current-loop bandwidth of ~10 kHz), it is perfectly reasonable to assume the current loop to be ideal at low frequencies around 15 Hz. Therefore, in the voltage-control block diagram shown in Figure 6.6a, the current closed-loop produces $\hat{I}_L$ equal to its reference value $\hat{I}_L^*$. In addition to a large DC component, $\hat{I}_L^*$ contains an unwanted second-harmonic frequency component $\hat{I}_{L2}$ due to v_{d2} in the input to the voltage controller. $\hat{I}_{L2}$ at the second-harmonic frequency results in a third-harmonic frequency distortion in the current drawn from the utility, as explained in Appendix 6A. Therefore, in the output of the voltage controller block in Figure 6.6a, $\hat{I}_{L2}$ is limited to approximately 1.5% of the DC component in $\hat{I}_L^*$.

The voltage control loop for low-frequency perturbations, in the range of the voltage-loop bandwidth of approximately 15 Hz, is shown in Figure 6.6b. As derived

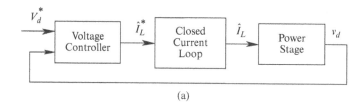

(a)

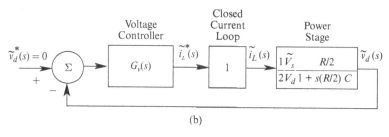

(b)

FIGURE 6.6   Voltage control loop.

in Appendix 6B, the transfer function of the power stage in Figure 6.6b at these low perturbation frequencies (ignoring the capacitor ESR) is:

$$\frac{\tilde{v}_d(s)}{\tilde{i}_L(s)} = \frac{1}{2}\frac{\hat{V}_s}{V_d}\frac{R/2}{1+s(R/2)C}. \tag{6.15}$$

To achieve a zero steady-state error, the voltage-controller transfer function should have a pole at the origin. However, since the PFC circuit is often a pre-regulator (not a strict regulator), this requirement is waived, which otherwise would make the voltage controller design much more complicated. The following simple transfer function is often used for the voltage controller in Figure 6.6b, where a pole is placed at the voltage-loop crossover frequency ω_{cv} (yet to be determined) below 15 Hz,

$$G_v(s) = \frac{k_v}{1+s/\omega_{cv}}. \tag{6.16}$$

At full load, the power stage transfer function given by Equation (6.15) has a pole at a very low frequency, for example, of the order of one or two Hz, which introduces a phase lag approaching 90° much beyond the frequency at which this pole occurs. The transfer function of the controller given by Equation (6.16) introduces a lag of 45° at the loop crossover frequency. Therefore, the sum of these two phase lags adds to ∼135° at the crossover frequency and results in a satisfactory phase margin of 45°. Using Equations (6.15) and (6.16), by definition, at the crossover frequency f_{cv}, the loop transfer function in Figure 6.6b has a magnitude equal to unity:

$$\left|\frac{k_v}{1+s/\omega_{cv}}\frac{1}{2}\frac{\hat{V}_s}{V_d}\frac{R/2}{1+s(R/2)C}\right|_{s=j(2\pi\times f_{cv})} = 1. \tag{6.17}$$

In the voltage controller of Figure 6.6b and Equation (6.16), the input $\hat{V}_{d2}$ results in an output $\hat{I}_{L2}$. Therefore, at the second-harmonic frequency in the voltage controller of Equation (6.16),

$$\left| \frac{k_v}{1+s/\omega_{cv}} \right|_{s=j(2\pi \times 120)} = \frac{\hat{I}_{L2}}{\hat{V}_{d2}}. \tag{6.18}$$

From Equations (6.17) and (6.18), the two unknowns k_v and ω_{cv} in the voltage controller transfer function of Equation (6.16) can be calculated, as described by a numerical example.

6.6 EXAMPLE OF SINGLE-PHASE PFC SYSTEMS

The operation and control of a PFC are demonstrated by means of an example, where the parameters are as shown in Table 6.1, and the total harmonic distortion in the input line current is required to be less than 3% [1, 2]:

6.6.1 Design of the Current-Control Loop

In Equation (6.9), assume that $\hat{V}_r = 1$. Following the procedure described in Chapter 4 for the peak-current-mode control of DC-DC converters, for the loop crossover frequency of 10 kHz($\omega_{ci} = 2\pi \times 10^4$ rad / s) and the phase margin of 60°, the parameters in the current controller of Equation (6.11) are as follows:

$$k_c = 4212$$
$$\omega_z = 1.68 \times 10^4 \text{ rad / s} \tag{6.19}$$
$$\omega_p = 2.35 \times 10^5 \text{ rad / s.}$$

Based on these parameter values given in Equation (6.19) of the transfer function $G_i(s)$ in Equation (6.11), the op-amp circuit is similar to that in Figure 4.19 in Chapter 4 with the following values for a chosen value of $R_1 = 100\,\text{k}\Omega$:

$$C_2 \simeq 0.17\,\text{nF}$$
$$C_1 \simeq 2.2\,\text{nF} \tag{6.20}$$
$$R_2 \simeq 27\,\text{k}\Omega.$$

TABLE 6.1 Parameters and operating values.

Nominal input AC source voltage, $V_{s,rms}$	120 V
Line frequency, f	60 Hz
Output voltage, V_d	250 V(DC)
Maximum power output	250 W
Switching frequency, f_s	100 kHz
Output filter capacitor, C	220 μF
ESR of the capacitor, r	100 $m\Omega$
Inductor, L_d	1 mH
Full-load equivalent resistance, R	250 Ω

6.6.2 Design of the Voltage-Control Loop

In this example, at full load, the plant transfer function given by Equation (6.15) has a pole at the frequency of 36.36 rad/s (5.79 Hz). At full load, $\hat{I}_L = 2.946\,\mathrm{A}$, and in Equation (6.8), $\hat{V}_{d2} = 6.029\,\mathrm{V}$. Based on the previous discussion, the second-harmonic component is limited to 1.5% of $\hat{I}_L$, such that $\hat{I}_{L2} = 0.0442\,\mathrm{A}$. Using these values, from Equations (6.17) and (6.18), the parameters in the voltage controller transfer function of Equation (6.16) are calculated: $k_v = 0.0754$, and $\omega_{cv} = 73.7\,\mathrm{rad/s}$ (11.73 Hz). This transfer function is realized by an op-amp circuit shown in Figure 6.7, where

$$R_1 = 100\,\mathrm{k\Omega}$$
$$R_2 = 7.54\,\mathrm{k\Omega} \tag{6.21}$$
$$C_1 = 1.8\,\mu\mathrm{F}.$$

6.7 SIMULATION RESULTS

The LTspice-based simulation of the PFC system is shown in Figure 6.8, where the input voltage and the full-bridge rectifier are combined for simplification purposes. The output load is decreased as a step at 100 ms. The resulting waveforms for the voltage and the inductor current are shown in Figure 6.9.

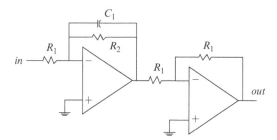

FIGURE 6.7 Op-amp circuit to implement transfer function $G_v(s)$.

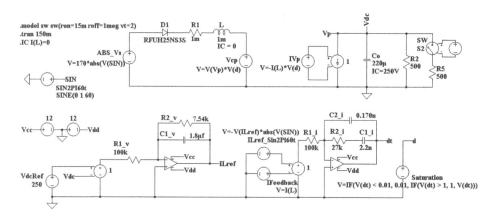

FIGURE 6.8 LTspice simulation diagram (the load is decreased at 100 ms); for a better resolution, execute the LTspice Schematic on the accompanying website.

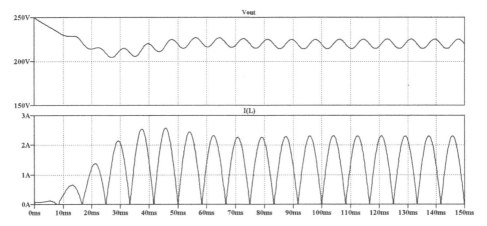

FIGURE 6.9 Simulation results: output voltage and inductor current.

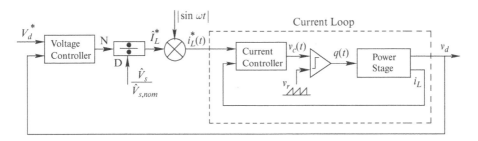

FIGURE 6.10 Feedforward of the input voltage.

6.8 FEEDFORWARD OF THE INPUT VOLTAGE

The input voltage is fed forward, as shown in Figure 6.10. In a system with a PFC interface, the output is nearly constant, independent of the changes in the RMS value of the input voltage from the utility. Therefore, an increase in the utility voltage $\hat{V}_s$ causes a decrease in $\hat{I}_L$, and vice versa. To avoid propagating the input voltage disturbance through the PFC feedback loops, the input voltage peak is fed forward, as shown in Figure 6.10, in determining $\hat{I}_L^*$.

6.9 OTHER CONTROL METHODS FOR PFCS

In this book, an average-current-control method is described for controlling PFCs. There are other methods, such as a single-cycle control. The reader is referred to References 4 and 5 for further exploration.

REFERENCES

1. Philip C. Todd, "UC3854 Controlled Power Factor Correction Circuit Design," Unitrode Application Note U-134.

2. Lyod Dixon, "High Power Factor Switching Pre-Regulator Design Optimization," Unitrode Design Application Manual.

PROBLEMS

6.1   In a single phase, 60-Hz power factor correction circuit, $V_s = 120\,\text{V(rms)}$, $V_d = 250\,\text{V}$, and the output power is 300 W. Calculate and draw the following waveforms, synchronized to v_s waveform: $\bar{i}_L(t)$, $d(t)$, and the average current $\bar{i}_d$ through the output diode.

6.2   In the numerical example given in this chapter, calculate the RMS input current if the utility voltage is $110\,\text{V(rms)}$, and compare it with its nominal value when $V_s = 120\,\text{V}\,(rms)$.

6.3   In problem 6.1, calculate the second-harmonic peak voltage in the capacitor if $C = 690\,\mu\text{F}$.

6.4   In the numerical example given in this chapter, calculate the maximum peak-peak ripple current in the inductor.

6.5   Repeat the design of the current loop in the given numerical example in this chapter if the loop crossover frequency is $25\,\text{kHz}$.

6.6   Repeat the design of the outer voltage loop in the numerical example given in this chapter if the output capacitance $C = 690\,\mu\text{F}$.

Simulation Problem

6.7   In the simulation diagram of Figure 6.8:
      (a) Observe the input voltage to the boost converter (output of the diode-rectifier bridge, not modeled here), the inductor current, and the voltage across the load.
      (b) List the harmonic components of the inductor current i_L and the capacitor current i_C.
      (c) Confirm the validity of Equation (6.5) for the current into the output stage of the PFC.

APPENDIX 6A Proof that $\dfrac{\hat{I}_{s3}}{\hat{I}_{L2}} = \dfrac{1}{2}$

The output of the voltage regulator G_v in Figure 6.6b in steady state is

$$\text{output of } G_v = \hat{I}_L - \hat{I}_{L,2}\cos 2\omega t, \tag{6A.1}$$

in which $\hat{I}_{L2}$, by a proper controller design is much smaller than $\hat{I}_L$, for example, less than 1.5%. The above expression is multiplied by $|\sin\omega t|$ to establish the reference for the inductor current. The second-harmonic distortion in Equation (6A.1) results in a third-harmonic distortion in the input AC current. This can be proven by multiplying the second-harmonic component in Equation (6A.1) with $\sin \omega t$, in order to see the distortion in the input AC current, as follows:

$$(-\hat{I}_{L2} \cos 2\omega t)\sin \omega t = \frac{1}{2}\hat{I}_{L2}\sin \omega t - \underbrace{\frac{1}{2}\hat{I}_{L2}\sin 3\omega t}_{\hat{i}_{s3}}. \tag{6A.2}$$

In Equation (6A.2), the fundamental-frequency component, due to the second-harmonic distortion, is compensated by the voltage-loop controller. However, the second-harmonic distortion with a peak $\hat{I}_{L2}$ results in a third-harmonic frequency distortion with one-half the amplitude. Therefore,

$$\frac{\hat{I}_{s3}}{\hat{I}_{L2}} = \frac{1}{2}. \tag{6A.3}$$

APPENDIX 6B Proof that $\dfrac{\tilde{v}_d}{\tilde{i}_L}(s) = \dfrac{1}{2}\dfrac{\hat{V}_s}{V_d}\dfrac{R/2}{1+s(R/2)C}$

In designing the controller, the output of $G_v(s)$ in Figure 6.6b under dynamic conditions has a strong DC component, a second harmonic frequency component i_{L2}, and a low-frequency (less than 15 Hz) perturbation term:

$$\text{Output of the voltage regulator} = \hat{I}_L + i_{L2} + \tilde{i}_L. \tag{6B.1}$$

Multiplying the reference current peak in Equation (6B.1) with $|\sin \omega t|$, the inductor current is

$$\bar{i}_L(t) = (\hat{I}_L + i_{L2} + \tilde{i}_L)|\sin \omega t|. \tag{6B.2}$$

In the circuit of Figure 6.2, assuming that the voltage drop across L_d is negligible,

$$|v_s(t)|\bar{i}_L(t) = (V_d + v_{d2} + \tilde{v}_d)(I_d + i_{d2} + \tilde{i}_d). \tag{6B.3}$$

Substituting into Equation (6B.3) $|v_s(t)| = \hat{V}_s|\sin \omega t|$ and the inductor current from Equation (6B.2),

$$\hat{V}_s|\sin \omega t|(\hat{I}_L + i_{L2} + \tilde{i}_L)|\sin \omega t| = (V_d + v_{d2} + \tilde{v}_d)(I_d + i_{d2} + \tilde{i}_d). \tag{6B.4}$$

Noting that $\left|\sin \omega t\right|^2 = \sin^2 \omega t = \dfrac{1}{2} - \dfrac{1}{2}\cos 2\omega t$ and neglecting the product of $\tilde{v}_d$ and $\tilde{i}_d$,

$$\frac{1}{2}\widehat{V}_s\widehat{I}_L + \frac{1}{2}\widehat{V}_s i_{L2} + \frac{1}{2}\widehat{V}_s\tilde{i}_L - \frac{\widehat{V}_s}{2}(\widehat{I}_L + i_{L2} + \tilde{i}_L)\cos(2\omega t)$$
$$\approx v_d(I_d + i_{d2}) + v_{d2}(I_d + I_{d2} + \tilde{i}_d) + \tilde{v}_d i_{d2} + \tilde{i}_d V_d + \tilde{v}_d I_d. \qquad (6B.5)$$

Equating the perturbation frequency terms on the two sides of the above equation,

$$\tilde{i}_d V_d + \tilde{v}_d I_d = \frac{1}{2}\widehat{V}_s\tilde{i}_L. \qquad (6B.6)$$

Recognizing that in the PFC of Figure 6.1a, $I_d = V_d / R$. Therefore in Equation (6B.6),

$$\left(\tilde{i}_d + \frac{\tilde{v}_d}{R}\right)V_d = \frac{1}{2}\widehat{V}_s\tilde{i}_L. \qquad (6B.7)$$

In the PFC output stage of Figure 6.1a, the voltage and the current are related as

$$\tilde{i}_d = \frac{1 + sRC}{R}\tilde{v}_d. \qquad (6B.8)$$

Substituting Equation (6B.8) into Equation (6B.7),

$$\frac{\tilde{v}_d}{\tilde{i}_L}(s) = \frac{1}{2}\frac{\widehat{V}_s}{V_d}\frac{R/2}{1 + s(R/2)C}. \qquad (6B.9)$$

7

MAGNETIC CIRCUIT CONCEPTS

The purpose of this chapter is to review some of the basic concepts associated with magnetic circuits and to develop an understanding of inductors and transformers needed in power electronics.

7.1 AMPERE-TURNS AND FLUX

Let us consider a simple magnetic structure of Figure 7.1 consisting of an N-turn coil with a current i, on a magnetic core made of iron. This coil applies Ni ampere-turns to the core. We will assume the magnetic field intensity H_m in the core to be uniform along the mean path length ℓ_m. The magnetic field intensity in the air gap is denoted as H_g. From Ampere's law, the closed line integral of the magnetic field intensity along the mean path within the core and in the air gap is equal to the applied ampere-turns:

$$H_m \ell_m + H_g \ell_g = Ni. \tag{7.1}$$

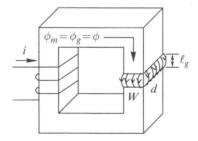

FIGURE 7.1 Magnetic structure with an air gap.

Power Electronics A First Course: Simulations and Laboratory Implementations, Second Edition.
Ned Mohan and Siddharth Raju.
© 2023 John Wiley & Sons, Inc. Published 2023 by John Wiley & Sons, Inc.
Companion Website: www.wiley.com/go/mohan/powerelectronics2e

In the core and in the air gap, the flux densities corresponding to H_m and H_g are as follows:

$$B_m = \mu_m H_m \tag{7.2}$$

$$B_g = \mu_o H_g, \tag{7.3}$$

where $\mu_o = 4\pi \times 10^{-7}\,\text{H}/\text{m}$. In terms of the above flux densities in Equation (7.1),

$$\frac{B_m}{\mu_m}\ell_m + \frac{B_g}{\mu_o}\ell_g = Ni. \tag{7.4}$$

Since flux lines form closed paths, the flux crossing any perpendicular cross-sectional area in the core is the same as that crossing the air gap. Therefore,

$$\phi = A_m B_m = A_g B_g \tag{7.5}$$

$$B_m = \frac{\phi}{A_m} \text{ and } B_g = \frac{\phi}{A_g}. \tag{7.6}$$

Substituting flux densities from Equation (7.6) into Equation (7.4),

$$\phi\left(\underbrace{\frac{\ell_m}{A_m \mu_m}}_{\mathfrak{R}_m} + \underbrace{\frac{\ell_g}{A_g \mu_o}}_{\mathfrak{R}_g}\right) = Ni \tag{7.7}$$

In Equation (7.7), the two terms within the parenthesis equal the reluctance $\mathfrak{R}_m$ of the core and the reluctance $\mathfrak{R}_g$ of the air gap, respectively. Therefore, the effective reluctance $\mathfrak{R}$ of the whole structure in the path of the flux lines is the sum of the two reluctances:

$$\mathfrak{R} = \mathfrak{R}_m + \mathfrak{R}_g. \tag{7.8}$$

Substituting Equation (7.8) into Equation (7.7),

$$\phi = \frac{Ni}{\mathfrak{R}}. \tag{7.9}$$

Equation (7.9) allows the flux ϕ to be calculated for the applied ampere-turns and hence B_m, and B_g can be calculated from Equation (7.6).

7.2 INDUCTANCE L

At any instant of time in the coil of Figure 7.2a, the flux linkage of the coil λ_m, due to flux lines entirely in the core, is equal to the flux ϕ_m times the number of turns N that

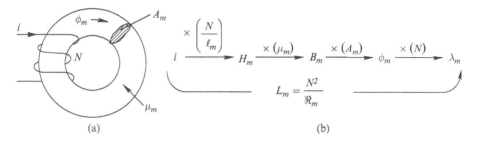

FIGURE 7.2 Coil inductance.

are linked. This flux linkage is related to the current i by a parameter defined as the inductance L_m:

$$\lambda_m = N\phi_m = L_m i, \tag{7.10}$$

where the inductance $L_m(=\lambda_m/i)$ is constant if the core material is in its linear operating region.

The coil inductance in the linear magnetic region can be calculated by multiplying all the factors shown in Figure 7.2b, which are based on earlier equations:

$$L_m = \frac{\lambda_m}{i} = \frac{\overbrace{\underbrace{\underbrace{\underbrace{\left(\dfrac{Ni}{\ell_m}\right)}_{H_m}\mu_m A_m}_{B_m} N}_{\phi_m}}}{i} = \frac{N^2}{\left(\dfrac{\ell_m}{\mu_m A_m}\right)} = \frac{N^2}{\mathcal{R}_m}. \tag{7.11}$$

Equation (7.11) indicates that the inductance L_m is strictly a property of the magnetic circuit (i.e. the core material, the geometry, and the number of turns), provided the operation is in the linear range of the magnetic material, where the slope of its *B-H* characteristic can be represented by a constant μ_m.

7.2.1 Energy Storage Due to Magnetic Fields

Energy in an inductor is stored in its magnetic field. From the study of electric circuits, we know that at any time, with a current i, the energy stored in the inductor is

$$W = \frac{1}{2} L_m i^2 [J], \tag{7.12}$$

where [J], for joules, is a unit of energy. Initially assuming a structure without an air gap, such as in Figure 7.2a, we can express the energy storage in terms of flux density, by substituting into Equation (7.12) the inductance from Equation (7.11), and the current from Ampere's law in Equation (7.1):

$$W_m = \frac{1}{2} \frac{N^2}{\underbrace{\frac{\ell_m}{\mu_m A_m}}} \underbrace{(H_m \ell_m / N)^2}_{i^2} = \frac{1}{2} \frac{(H_m \ell_m)^2}{\underbrace{\frac{\ell_m}{\mu_m A_m}}} = \frac{1}{2} \frac{B_m^2}{\mu_m} \underbrace{A_m \ell_m}_{volume} [J], \qquad (7.13)$$

where $A_m \ell_m = volume$, and in the linear region $B_m = \mu_m H_m$. Therefore, from Equation (7.13), the energy density in the core is

$$w_m = \frac{1}{2} \frac{B_m^2}{\mu_m}. \qquad (7.14)$$

Similarly, the energy density in the air gap depends on μ_o and the flux density in it. Therefore, from Equation (7.14), the energy density in any medium can be expressed as

$$w = \frac{1}{2} \frac{B^2}{\mu} [J / m^3]. \qquad (7.15)$$

In inductors, the energy is primarily stored in the air gap purposely introduced in the path of flux lines.

In transformers, there is no air gap in the path of the flux lines. Therefore, the energy stored in the core of an ideal transformer is zero, where the core permeability is assumed infinite, and hence H_m is zero for a finite flux density. In a real transformer, the core permeability is finite, resulting in some energy storage in the core.

7.3 FARADAY'S LAW: INDUCED VOLTAGE IN A COIL DUE TO TIME-RATE OF CHANGE OF FLUX LINKAGE

In our discussion so far, we have established in magnetic circuits relationships between the electrical quantity i and the magnetic quantities H, B, ϕ, and λ. These relationships are valid under DC (static) conditions, as well as at any instant when these quantities vary with time. We will now examine the voltage across the coil under time-*varying* conditions. In the coil of Figure 7.3, Faraday's law dictates that the time-rate of change of flux-linkage equals the voltage across the coil at any instant:

$$e(t) = \frac{d}{dt}\lambda(t) = N \frac{d}{dt}\phi(t). \qquad (7.16)$$

This assumes that all flux lines link all N-turns such that $\lambda = N\phi$. The polarity of the emf $e(t)$ and the direction of $\phi(t)$ in the above equation are yet to be justified.

The relationship in Equation (7.16) is valid, no matter what is causing the flux to change. One possibility is that a second coil is placed on the same core. When the second coil is supplied by a time-varying current, mutual coupling causes the flux ϕ through the coil to change with time. The other possibility is that a voltage $e(t)$ is applied across the coil in Figure 7.3, causing the change in flux, which can be calculated by integrating both sides of Equation (7.16) with respect to time:

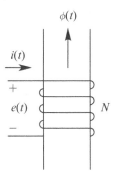

FIGURE 7.3 Voltage polarity and direction of flux and current.

$$\phi(t) = \phi(0) + \frac{1}{N} \int_0^t e(\tau) \cdot d\tau, \tag{7.17}$$

where $\phi(0)$ is the initial flux at $t = 0$ and τ is a variable of integration.

Recalling Ohm's law, $v = Ri$, the current direction through a resistor is into the terminal at the positive polarity. This is the passive sign convention. Similarly, in the coil of Figure 7.3, we can establish the voltage polarity and the flux direction in order to apply Faraday's law, given by Equations (7.16) and (7.17). If the flux direction is given, we can establish the voltage polarity as follows: first, determine the direction of a hypothetical current that will produce flux in the same direction as given. Then, the positive polarity for the voltage is at the terminal, which this hypothetical current is entering. Conversely, if the voltage polarity is given, imagine a hypothetical current entering the positive-polarity terminal. This current, based on how the coil is wound, for example, in Figure 7.3, determines the flux direction for use in Equations (7.16) and (7.17). Following these rules to determine the voltage polarity and the flux direction is easier than applying Lenz's law (not discussed here).

The voltage is induced due to $d\lambda / dt$, regardless of whether any current flows in that coil.

7.4 LEAKAGE AND MAGNETIZING INDUCTANCES

Just as conductors guide currents in electric circuits, magnetic cores guide *flux* in *magnetic circuits*. But there is an important difference. In electric circuits, the conductivity of copper is approximately 10^{20} times higher than that of air, allowing leakage currents to be neglected at DC or at low frequencies such as 60 Hz. In magnetic circuits, however, the permeabilities of magnetic materials are, at best, only 10^4 times greater than that of air. Because of this relatively low ratio, the core window in the structure of Figure 7.4a has "leakage" flux lines, which do not reach their intended destination, which may be, for example, another winding in a transformer or an air gap in an inductor. Note that the coil shown in Figure 7.4a is drawn schematically. In practice, the coil consists of multiple layers, and the core is designed to fit as snugly to the coil as possible, thus minimizing the unused "window" area.

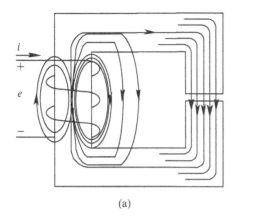 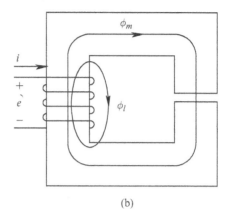

(a)   (b)

FIGURE 7.4 (a) Magnetic and leakage fluxes; (b) equivalent representation of magnetic and leakage fluxes.

The leakage effect makes accurate analysis of magnetic circuits more difficult, requiring sophisticated numerical methods, such as finite element analysis. However, we can account for the effect of leakage fluxes by making certain approximations. We can divide the total flux ϕ into two parts:

1. The magnetic flux ϕ_m, which is completely confined to the core and links all N turns, and,
2. The leakage flux, which is partially or entirely in air and is represented by an "equivalent" leakage flux ϕ_ℓ, which also links all N turns of the coil but does not follow the entire magnetic path, as shown in Figure 7.4b.

In Figure 7.4b, $\phi = \phi_m + \phi_\ell$, where ϕ is the equivalent flux that links all N turns. Therefore, the total flux linkage of the coil is

$$\lambda = N\phi = \underbrace{N\phi_m}_{\lambda_m} + \underbrace{N\phi_\ell}_{\lambda_\ell} = \lambda_m + \lambda_\ell. \tag{7.18}$$

The total inductance (called the self-inductance) can be obtained by dividing both sides of Equation (7.18) by the current i:

$$\underbrace{\frac{\lambda}{i}}_{L_{self}} = \underbrace{\frac{\lambda_m}{i}}_{L_m} + \underbrace{\frac{\lambda_\ell}{i}}_{L_\ell} \tag{7.19}$$

$\therefore$
$$L_{self} = L_m + L_\ell, \tag{7.20}$$

where L_m is often called the *magnetizing inductance* due to ϕ_m in the magnetic core, and L_ℓ is called the *leakage inductance* due to the leakage flux ϕ_ℓ. From Equations (7.19) and (7.20), the total flux linkage of the coil in Equation (7.18) can be written as

$$\lambda = (L_m + L_\ell)i. \tag{7.21}$$

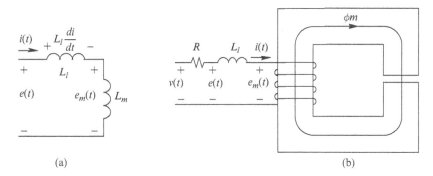

FIGURE 7.5 (a) Circuit representation; (b) leakage inductance separated from the core.

Hence, from Faraday's law in Equation (7.16),

$$e(t) = \underbrace{L_m \frac{di}{dt}}_{e_m(t)} + L_\ell \frac{di}{dt}. \tag{7.22}$$

This results in the electrical circuit of Figure 7.5a. In Figure 7.5b, the voltage drop due to the leakage inductance can be shown separately so that the voltage induced in the coil is solely due to the magnetizing flux. The coil resistance R can then be added in series to complete the representation of the coil.

7.4.1 Mutual Inductances

Most magnetic circuits, such as those encountered in inductors and transformers, consist of multiple coils. In such circuits, the flux established by the current in one coil partially links the other coil or coils. This phenomenon can be described mathematically by means of mutual inductances, as examined in circuit theory courses. However, we will use simpler and more intuitive means to analyze mutually coupled coils, as in a flyback converter discussed in Chapter 8 dealing with transformer-isolated DC-DC converters.

7.5 TRANSFORMERS

In power electronics, high-frequency transformers are essential to switch-mode DC power supplies. Such transformers often consist of two or more tightly coupled windings where almost all of the flux produced by one winding links the other windings. Including the leakage flux in detail makes the analysis very complicated and not very useful for our purposes here. Therefore, we will include only the magnetizing flux ϕ_m that links all the windings, ignoring the leakage flux, whose consequences will be acknowledged separately.

To understand the operating principles of transformers, we will consider a three-winding transformer, shown in Figure 7.6, such that this analysis can be extended to any number of windings.

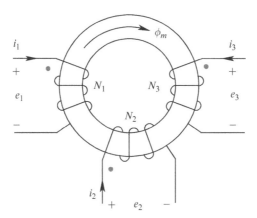

FIGURE 7.6   Transformer with three windings.

Faraday's law: In this transformer, all windings are linked by the same flux ϕ_m. Therefore, from Faraday's law, the induced voltages at the dotted terminals with respect to their undotted terminals are as follows:

$$e_1 = N_1 \frac{d\phi_m}{dt} \tag{7.23}$$

$$e_2 = N_2 \frac{d\phi_m}{dt} \tag{7.24}$$

$$e_3 = N_3 \frac{d\phi_m}{dt}. \tag{7.25}$$

The above equations based on Faraday's law result in the following relationship that shows that the volts-per-turn induced in each winding is the same due to the same rate of change of flux that links them,

$$\frac{d\phi_m}{dt} = \frac{e_1}{N_1} = \frac{e_2}{N_2} = \frac{e_3}{N_3}. \tag{7.26}$$

Equation (7.26) shows how desired voltage-ratios between various windings can be achieved by selecting the appropriate winding turns ratios. The instantaneous flux ϕ_m is obtained by expressing Equation (7.26) in its integral form below with proper integral limits,

$$\phi_m = \frac{1}{N_1} \int e_1 dt = \frac{1}{N_2} \int e_2 dt = \frac{1}{N_3} \int e_3 dt. \tag{7.27}$$

Ampere's law: In accordance with Ampere's law given in Equation (7.9), the flux ϕ_m at any instant of time is supported by the net magnetizing ampere-turns applied to the core in Figure 7.6,

$$N_1 i_1 + N_2 i_2 + N_3 i_3 = \Re_m \phi_m. \tag{7.28}$$

In Equation (7.28), $\Re_m$ is the reluctance in the flux path of the core of Figure 7.6, and the currents are defined as positive into the dotted terminals of each winding such as to produce flux lines in the same direction. The net ampere-turns consist of various winding currents that depend on the circuits connected to them.

Equations (7.27) and (7.28) are the key to understanding transformers: to one of the windings, the applied voltage, equal to the induced voltage in it if the winding resistance and the leakage flux are ignored, results in flux ϕ_m which is supported by the net magnetizing ampere-turns given by Equation (7.28), overcoming the core reluctance.

7.5.1 Transformer Equivalent Circuit

It is often useful to have an equivalent circuit of a transformer such as that shown in Figure 7.7b. Before developing this equivalent circuit, consider this to be an ideal transformer, with an infinite core permeability resulting in $\Re_m = 0$. Therefore, the net magnetizing ampere-turns in Equation (7.28) are zero, and such an ideal transformer is shown in Figure 7.7a.

A practical transformer such as that in Figure 7.6 doesn't have infinite core permeability and hence needs net ampere-turns to support the core flux. Although any of the windings could have been selected, let us select winding 1 to deliver the net magnetizing ampere-turns, with a magnetizing current i_{m1} flowing through N_1 turns. Therefore, in Equation (7.28),

$$N_1 i_{m1} = \Re_m \phi_m. \tag{7.29}$$

From Equations (7.28) and (7.29), we can write the following,

$$N_1 \underbrace{(i_1 - i_{m1})}_{i_1'} + N_2 i_2 + N_3 i_3 = 0, \tag{7.30}$$

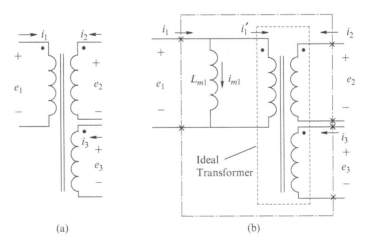

(a)　　　　　　　　(b)

FIGURE 7.7 Equivalent circuits of transformers: (a) ideal and (b) actual (leakage impedances not shown).

where i_1 can be considered as consisting of the sum of two components,

$$i_1 = i_m + i_1',\tag{7.31}$$

and, from Equation (7.30),

$$N_1 i_1' + N_2 i_2 + N_3 i_3 = 0.\tag{7.32}$$

The net ampere-turns in Equation (7.32) equal zero, and hence this equation corresponds to the ideal-transformer portion of the equivalent circuit, as shown in Figure 7.7b. In i_1 of Equation (7.31) and Figure 7.7b, the magnetizing current i_{m1} flows through the magnetizing inductance L_{m1}, as justified on the following page.

From Equation (7.29),

$$\phi_m = \frac{N_1 i_{m1}}{\Re_m}.\tag{7.33}$$

Substituting for ϕ_m from Equation (7.33) into Equation (7.23),

$$e_1 = \left(\frac{N_1^2}{\Re_m}\right)\frac{di_{m1}}{dt},\tag{7.34}$$

where, using Equation (7.11), the quantity within the brackets in the equation above is the magnetizing inductance of winding 1,

$$L_{m1} = \frac{N_1^2}{\Re_m}.\tag{7.35}$$

The analysis above is based on neglecting the leakage flux, assuming that the flux produced by a winding links all the other windings. In a simplified analysis, the leakage flux of a winding can be assumed to result in a leakage inductance, which can be added, along with the winding resistance, in series with the induced voltage $e(t)$ in the winding in the equivalent circuit representation. A systematic description of the principle on which transformers operate is presented in [1].

REFERENCE

1. N. Mohan, T.M. Undeland, and W.P. Robbins, *Power Electronics: Converters, Applications and Design*, 3rd Edition (New York: John Wiley & Sons, 2003).

PROBLEMS

Inductors

The magnetic core in Problems 7.1 through 7.5 has the following properties: the core area $A_m = 0.931\,\mathrm{cm}^2$, the magnetic path length of $\ell_m = 3.76\,\mathrm{cm}$, and the relative permeability of the material is $\mu_r = \mu_m/\mu_o = 5000$.

7.1 Calculate the reluctance $\Re$ of this core.

7.2 Calculate the reluctance of an air gap of length $\ell_g = 1\,\text{mm}$ if it is introduced in the core of Problem 7.1.

7.3 A coil with $N = 25$ turns is wound on a core with an air gap described in Problem 7.2. Calculate the inductance of this coil.

7.4 If the flux density in the core in Problem 7.3 is not to exceed $0.2\,\text{T}$, what is the maximum current that can be allowed to flow through this inductor coil?

7.5 At the maximum current calculated in Problem 7.4, calculate the energy stored in the magnetic core and the air gap, and compare the two.

Two-Winding Transformers

7.6 In Figure 7.7, assume that the transformer has only two windings: winding 1 and winding 2, with $N_1 / N_2 = 5$ and the leakage inductances are to be neglected. Winding 1 is applied a voltage $v_1 = \sqrt{2} \times 120 \sin \omega t$ volts at a frequency of $20\,\text{kHz}$. The load at the second winding is a resistor $R_L = 10\,\Omega$. Assuming this to be an ideal transformer so that the magnetizing current is zero, calculate and plot v_1, i_1 and i_2. What would this plot look like if i_2 is defined to be coming out of the dotted terminal of winding 2.

7.7 Repeat Problem 7.6, if the rms value of the magnetizing current is 1/5 of the RMS value of the winding-1 current in Problem 7.6.

Three-Winding Transformers

In Problems 7.8 through 7.13, the three-winding transformer with $N_1 = 10$ turns, $N_2 = 5$ turns, and $N_3 = 5$ turns uses a magnetic core that has the following properties: $A_m = 0.639\,\text{cm}^2$, the magnetic path length of $\ell_m = 3.12\,\text{cm}$, and the relative permeability of the material $\mu_r = \mu_m / \mu_o = 5000$. A square-wave voltage, of 30-V amplitude, alternating between $-30\,\text{V}$ and $+30\,\text{V}$, at a frequency of 100 kHz, is applied to winding 1. Ignore the leakage inductances and assume the flux waveform to be symmetric with the same positive and negative peak amplitudes.

7.8 Calculate and draw the magnetizing current waveform, along with the applied voltage waveform, and the waveforms of the voltages induced in windings 2 and 3, assuming them to be open.

7.9 Calculate the self-inductances of each winding in Problem 7.8.

7.10 Calculate the peak flux density in Problem 7.8.

7.11 A load resistance of $10\,\Omega$ is connected to winding 2, and winding 3 is open. Calculate and draw the currents in windings 1 and 2, along with the applied voltage waveform.

7.12 Assuming that winding 1 is not applied a voltage, and windings 1 and 2 are open, what is the peak amplitude of the square-wave voltage at 100 kHz that can be applied to winding 3 of this transformer, if the peak flux density calculated in Problem 7.10 is not to be exceeded?

7.13 In Problem 7.8, what is the peak amplitude of the voltage that can be applied to winding 1 without exceeding the peak flux density calculated in Problem 7.10, if the frequency of the square wave voltage is 200 kHz? What is the peak value of the magnetizing current, as compared to the one at 100 kHz?

8
SWITCH-MODE DC POWER SUPPLIES

8.1 APPLICATIONS OF SWITCH-MODE DC POWER SUPPLIES

Switch-mode DC power supplies represent an important power electronics application area with a worldwide market of several billion dollars per year. Many of these power supplies incorporate transformer isolation for reasons that are discussed below. Within these power supplies, transformer-isolated DC-DC converters are derived from non-isolated DC-DC converter topologies already discussed in Chapter 3. For short, we will refer to transformer-isolated switch-mode DC power supplies as SMPS (switched-mode power supply, sometimes called switcher), whose block diagram is shown in Figure 8.1. As shown in Figure 8.1, these supplies encompass the rectification of the utility supply, and the voltage V_{in} across a large filter capacitor is the input to the transformer-isolated DC-DC converter, which is the focus of discussion in this chapter. Internally, the transformer operates at very high frequencies, typically upwards of a few hundred kHz, thus resulting in small size and weight, as discussed in the next chapter.

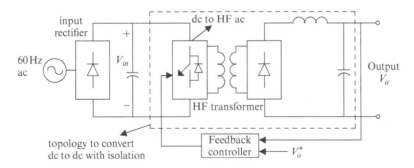

FIGURE 8.1 Block diagram of switch-mode DC power supplies.

Power Electronics A First Course: Simulations and Laboratory Implementations, Second Edition.
Ned Mohan and Siddharth Raju.
© 2023 John Wiley & Sons, Inc. Published 2023 by John Wiley & Sons, Inc.
Companion Website: www.wiley.com/go/mohan/powerelectronics2e

8.2   NEED FOR ELECTRICAL ISOLATION

Electrical isolation by means of transformers is needed in switch-mode DC power supplies for three reasons:

1. Safety: It is necessary for the low-voltage DC output to be isolated from the utility supply to avoid the shock hazard.
2. Different reference potentials: The DC supply may have to operate at a different potential, for example, the DC supply to the gate drive for the upper MOSFET in the power pole is referenced to its source.
3. Voltage matching: If the DC-DC conversion is large, then to avoid requiring large voltage and current ratings of semiconductor devices, it may be economical and operationally more suitable to use an electrical transformer for conversion of voltage levels.

8.3   CLASSIFICATION OF TRANSFORMER-ISOLATED DC-DC CONVERTERS

In the block diagram of Figure 8.1, there are the following three categories of transformer-isolated DC-DC converters, all of which are discussed in detail in this chapter:

1. Flyback converters derived from buck-boost DC-DC converters
2. Forward converter derived from buck DC-DC converters
3. Full-bridge and half-bridge converters derived from buck DC-DC converters

8.4   FLYBACK CONVERTERS

Flyback converters are very commonly used in applications at low power levels below 50 W. These are derived from the buck-boost converter redrawn in Figure 8.2a, where the inductor is drawn descriptively on a low permeability core.

The flyback converter in Figure 8.2b consists of two mutually coupled coils, where the coil orientations are such that at the instant when the transistor is turned off, the current switches to the second coil to maintain the same flux in the core. Therefore, the dots on coils are as shown in Figure 8.2b, where the current into the dot of either coil produces core flux in the same direction. Commonly, the circuit of Figure 8.2b is redrawn as in Figure 8.2c.

We will consider the steady state in the incomplete demagnetization mode, where the energy is never completely depleted from the magnetic core. This corresponds to the continuous conduction mode (CCM) in buck-boost converters. We will assume ideal devices and components, the output voltage $v_o(t) = V_o$, and the leakage inductances to be zero.

Turning on the transistor at $t = 0$ in the circuit in Figure 8.2c applies the input voltage V_{in} across coil 1, and the core magnetizing flux ϕ_m increases linearly from its initial value $\phi_m(0)$, as shown in the waveforms of Figure 8.3. During the transistor on interval DT_s, the increase in flux can be calculated from Faraday's law as:

$$\Delta\phi_{p-p} = \frac{V_{in}}{N_1} DT_s \tag{8.1}$$

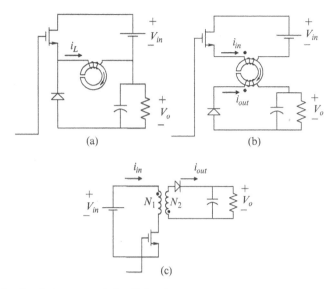

FIGURE 8.2 Buck-boost and the flyback converters.

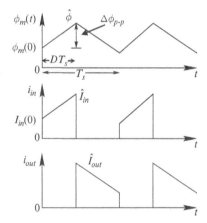

FIGURE 8.3 Flyback converter waveforms.

Due to increasing ϕ_m, the induced voltage $(N_2 / N_1)V_{in}$ across coil 2 adds to the output voltage V_o to reverse bias the diode, resulting in $i_{out} = 0$. Corresponding to the core flux, the current i_{in} can be calculated using the relationship $\phi = Ni / \Re$, where $\Re$ is the core reluctance in the flux path. Therefore, using Equation (8.1), the increase in the input current during the on interval from its initial value $I_{in}(0)$ can be calculated as

$$\Delta i_{in} = \frac{\Re}{N_1^2} V_{in} D T_s \tag{8.2}$$

During the on interval DT_s, the output load is entirely supplied by the energy stored in the output capacitor, and the core magnetizing flux and the input current reach their peak values at the end of this interval:

$$\hat{\phi}_m = \phi_m(0) + \frac{V_{in}}{N_1} DT_s \tag{8.3}$$

$$\hat{I}_{in} = I_{in}(0) + \frac{\Re}{N_1^2} V_{in} DT_s = I_{in}(0) + \frac{V_{in}(DT_s)}{L_{m1}}, \tag{8.4}$$

where $L_{m1}(= N_1^2/\Re)$ is the magnetizing inductance of the transformer seen from the primary side. After the on interval, turning off the transistor forces the input current in Figure 8.2c to zero. The magnetic energy stored in the magnetic core due to the flux ϕ_m cannot change instantaneously, and hence the ampere-turns applied to the core must be the same at the instant immediately before and after turning the transistor off. Therefore, the current i_{out} in coil 2 through the diode suddenly jumps to its peak value such that

$$N_2 \hat{I}_{out}\big|_{i_{in}=0} = N_1 \hat{I}_{in}\big|_{i_{out}=0} \tag{8.5}$$

$$\therefore \qquad\qquad \hat{I}_{out} = \frac{N_1}{N_2} \hat{I}_{in}. \tag{8.6}$$

With the diode conducting, the output voltage V_o appears across coil 2 with a negative polarity. Hence, during the off interval $(1-D)T_s$, the core flux declines linearly, as plotted in Figure 8.3, by $\Delta\phi_{p-p}$, where

$$\Delta\phi_{p-p} = \frac{V_o}{N_2}(1-D)T_s. \tag{8.7}$$

Using Equations (8.1) and (8.7),

$$\frac{V_o}{V_{in}} = \left(\frac{N_2}{N_1}\right)\frac{D}{1-D}. \tag{8.8}$$

The change in the current $i_{out}(t)$ can be calculated in a manner similar to Equation (8.2), and this current is plotted in Figure 8.3.

Equation (8.8) shows that in a flyback converter, the dependence of the voltage ratio on the duty ratio D is identical to that in the buck-boost converter, and it also depends on the coils' turns ratio N_2/N_1. Flyback converters require the minimum number of components by integrating the inductor (needed for a buck-boost operation) with the transformer that provides electrical isolation and matching of the voltage levels. These converters are very commonly used in low-power applications in the complete demagnetization mode (corresponding to the discontinuous conduction mode in buck-boost), which makes their control easier. A disadvantage of flyback converters is the need for snubbers to prevent voltage spikes across the transistor and diode due to leakage inductances associated with the two coils.

Example 8.1
In a flyback converter, shown in Figure 8.2c, $V_{in} = 48\,V$, $V_o = 5\,V$, $N_1/N_2 = 6$, and the magnetizing inductance $L_{m1} = 150\,\mu H$. This converter is operating in equivalent CCM with a switching frequency $f_s = 200\,kHz$ and supplying an output load $P_o = 60\,W$. Assuming this converter to be lossless, calculate the waveforms associated with it.

Solution From Equation (8.8), the duty ratio $D = 0.385$, where $T_s = 5\,\mu s$. The average currents are $I_{in} = 0.625\,A$ and $I_{out} = 6\,A$. In Figure 8.3, the rise in current during the on interval DT_s can be calculated as

$$\hat{I}_{in} - I_{in}(0) = \frac{V_{in}(DT_s)}{L_{m1}} = 0.616\,A$$

From the waveforms of Figure 8.3, the average input current can be calculated as follows:

$$I_{in} = \frac{\hat{I}_{in} + I_{in}(0)}{2} D = 0.625\,A\,; \qquad \hat{I}_{in} + I_{in}(0) = 3.247\,A\,.$$

From the equations above, in Figure 8.3, $\hat{I}_{in} = 1.93\,A$ and $I_{in}(0) = 1.315\,A$. The output current has a peak value $\hat{I}_{out} = \hat{I}_{in} \frac{N_1}{N_2} = 11.58\,A$ and $I_{out}(0) = I_{in}(0) \frac{N_1}{N_2} = 7.89\,A$.

8.4.1 Simulation and Hardware Prototyping—CCM without Snubber

The simulation of a nonideal flyback converter is demonstrated by means of an example:

Example 8.2
In the flyback converter shown in Figure 8.2c, $C = 490\,\mu F$, and $R = 10\,\Omega$. It is operating in DC steady state under the following conditions: $V_{in} = 15\,V$, $D = 0.5$, and $f_s = 100\,kHz$. The transformer's primary-to-secondary turns ratio is 2:1. The primary side has a magnetizing inductance of $L_m = 27.8\,\mu H$ and a leakage inductance of $L_l = 0.26\,\mu H$. For the switch and the diode, use the parameters given in the Appendix of Chapter 2. Simulate this converter using LTspice.

Solution The simulation file used in this example is available on the accompanying website. The LTspice model is as shown in Figure 8.4, and the steady-state waveforms from the simulation of this model are shown in Figure 8.5.

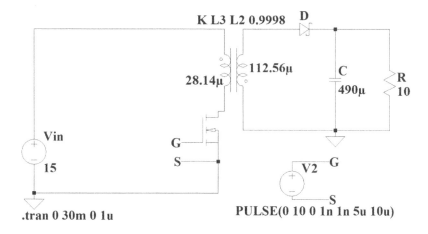

FIGURE 8.4 LTspice model.

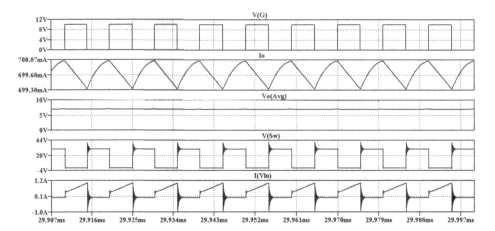

FIGURE 8.5   LTspice simulation results.

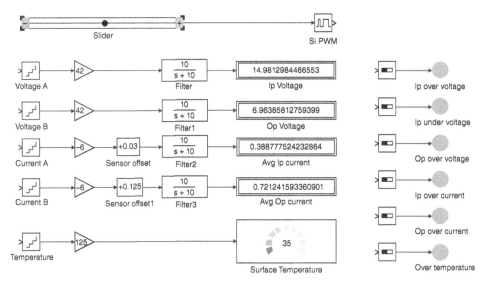

FIGURE 8.6   Workbench model.

The Workbench model for implementing the above example in hardware using the Sciamble lab kit is as shown in Figure 8.6.

The steady-state waveforms from running the flyback converter using the Sciamble laboratory kit are shown in Figure 8.7. The step-by-step procedure for re-creating the above hardware implementation is presented in [1].

8.4.2   RCD Snubber

Unlike the case of an ideal flyback converter waveform shown in Figure 8.3, the switch-node voltage waveform of a practical flyback converter, shown in Figure 8.8, has significant ringing. The energy stored in the leakage inductance L_l, unlike the

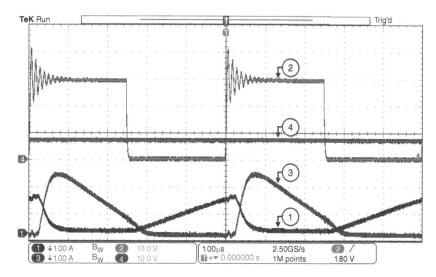

FIGURE 8.7 Workbench hardware results: (1) Input current, (2) Switch-node voltage, (3) Diode current and (4) Output voltage.

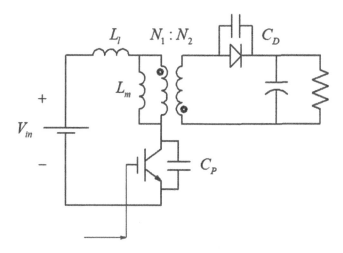

FIGURE 8.8 Practical flyback converter parasitic elements.

energy stored in the magnetizing inductance L_m, does not transfer over to the secondary side when the switch is turned off. This energy charges up the switch's parasitic capacitance, leading to a high voltage across the switch, causing significant voltage stress, which can damage the switch, the transformer insulation, or both.

In a practical flyback converter, the magnitude of the voltage swing is clamped by the use of a snubber circuit [2]. This section presents the operation and design of one of the commonly used snubber circuit, RCD snubber, shown in Figure 8.9.

We will first analyze the steady-state operation of the snubber circuit before proceeding with how to determine the snubber parameters. The voltage across the snubber capacitor is assumed to be a constant V_{cs}, which is justified given a sufficiently large C_s.

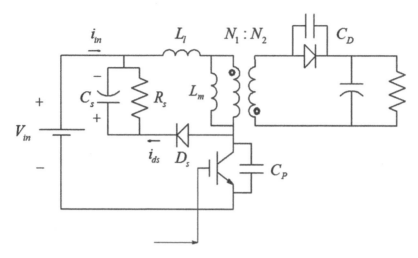

FIGURE 8.9 RCD snubber.

8.4.2.1 Steady-State Operation of RCD Snubber

At time $0 < t < DT_s$, when the transistor is on, as shown in Figure 8.10a, the primary current increases linearly, as shown in Figure 8.10b. During this period, the voltage across the transistor is 0 as shown in Figure 8.10c, and there is no current flowing into the snubber circuit because the diode D_s is reverse biased with a voltage of $V_{in} + V_{cs}$ across it, as shown in Figure 8.10d. The snubber capacitor is slowly discharged by the snubber resistor, as shown by the negative current, $-\frac{V_{cs}}{R_s}$ in Figure 8.10e.

When the transistor is turned off, the parasitic capacitance C_P is charged up to the sum of the input voltage and the reflected secondary voltage, $V_{in} + nV_o$ where $n = \frac{N_1}{N_2}$. During this interval, $DT_s < t < T_1$, the energy stored in the magnetizing inductance contributes to a very rapid charging of the transistor's parasitic capacitance C_P as well as the discharging of the secondary side diode's junction capacitance C_D. The input current i_{in} remains relatively constant, given the negligible energy transfer from the magnetizing inductance to the parasitic capacitances.

Once the secondary side diode's junction capacitance is fully discharged, and the diode begins conducting and simultaneously C_P is charged to $V_{in} + nV_o$, the energy stored in the magnetizing inductance starts charging the output capacitor.

During the time interval $T_1 < t < T_2$, the snubber diode is still reverse biased since $nV_o < V_{Cs}$. The voltage across the transistor continues to rise until it reaches $V_{in} + V_{Cs}$ (plus the diode forward voltage drop, which is assumed to be negligible in this discussion), forward biasing the snubber diode. The primary current during this interval falls rapidly as only the energy stored in the primary side leakage inductance is available to charge C_P.

In the absence of the snubber circuit, the energy in the leakage inductance continues to charge C_P, which leads to high voltage ringing across the transistor, as seen in Figure 8.7. In addition to the issue of high voltage, the ringing frequency, given by Equation (8.9), is typically in the order of MHz and is a source of significant noise and EMI:

$$f_p = \frac{1}{2\pi\sqrt{L_l C_p}}. \tag{8.9}$$

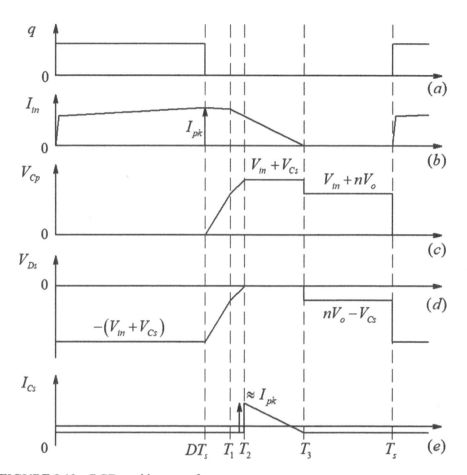

FIGURE 8.10 RCD snubber waveforms.

In the presence of the snubber circuit, the primary current flows through the forward-biased snubber diode, charging the snubber capacitor. V_{Cs} remains relatively a constant by design, as explained in the next section. During this interval, $T_2 < t < T_3$, the primary current continues to fall rapidly. The voltage across the transistor remains clamped at $V_{in} + V_{Cs}$.

Once the leakage energy is fully transferred to the clamp capacitor at time $t = T_3$, the primary current goes to 0, and the snubber diode gets reverse biased as the transistor voltage drops from $V_{in} + V_{Cs}$ to $V_{in} + nV_o$.

Example 8.3

In Example 8.1, calculate the frequency of the ringing of the voltage across the transistor when it turns off. The transistor's parasitic capacitance $C_P = 970\,pF$.

Solution　From Equation (8.9), $f_p = \frac{1}{2\pi\sqrt{L_l C_p}} \approx 10\,\text{MHz}$.

8.4.2.2　Design of RCD Snubber

At steady state, since the voltage across the snubber capacitor is a constant, the current through the capacitor averaged over a switching cycle $T_s\left(= \frac{1}{f_s}\right)$ is zero:

$$\frac{1}{2} I_{pk} (T_3 - T_2) - \frac{V_{Cs}}{R_s} T_s = 0. \tag{8.10}$$

During the time interval $T_2 < t < T_3$, when the snubber diode is forward biased, the voltage across the inductor from Figure 8.8 is the sum of the voltage across the snubber capacitor and the voltage across the transformer primary winding:

$$V_{Ll} = L_l \frac{\Delta I_l}{\Delta t} = V_{Cs} - nV_o. \tag{8.11}$$

The current through the leakage inductor goes from approximately I_{pk} to zero, as seen in Figure 8.10b. The duration of the interval when the snubber diode is conducting can be obtained from Equation (8.11):

$$L_l \frac{\Delta I_l}{\Delta t} = L_l \frac{I_{pk} - 0}{T_3 - T_2} = V_{Cs} - nV_o \Rightarrow T_{sn} = T_3 - T_2 = L_l \frac{I_{pk}}{V_{Cs} - nV_o}. \tag{8.12}$$

The expression for the snubber resistor can be obtained using Equations (8.10) and (8.12):

$$\frac{1}{2} I_{pk}^2 L_l \frac{1}{V_{Cs} - nV_o} - \frac{V_{Cs}}{R_s} T_s = 0$$

$$\Rightarrow R_s = \frac{V_{Cs} (V_{Cs} - nV_o)}{\frac{1}{2} I_{pk}^2 L_l f_s}. \tag{8.13}$$

Given the converter operating condition and the desired maximum transistor voltage, the snubber resistor can be determined using the above expression. The lower the tolerable voltage across the transistor, the smaller the snubber resistor and thus higher losses in the snubber circuit. The power lost in the snubber circuit is:

$$P_s = \frac{V_{Cs}^2}{R_s} = \frac{1}{2} I_{pk}^2 L_l f_s \frac{V_{Cs}}{(V_{Cs} - nV_o)} = \alpha P_{Ll}, \tag{8.14}$$

where $P_{Ll} = \frac{1}{2} I_{pk}^2 L_l f$ is the leakage inductor power, the energy stored in the leakage inductor per switching cycle time period. $\alpha = 1 / \left(1 - \frac{nV_o}{V_{Cs}}\right)$ is the factor by which the losses in the circuit are scaled due to the presence of the snubber compared to the circuit without a snubber. Since $V_{Cs} > nV_o$, α is always greater than 1. The higher the snubber voltage, the lower the losses and vice versa.

The value of the snubber capacitor isn't as crucial as that of the resistor. The only requirement is that the capacitance is large enough to maintain the snubber voltage a constant. An approximate expression can be obtained from the snubber capacitor current waveform in Figure 8.10e. Assuming that the current in the snubber resistor is negligible compared to the peak input current I_{pk}, during the interval when the snubber diode is reverse biased:

$$i_{Cs} = C_s \frac{\Delta V_{Cs}}{\Delta T} \Rightarrow \frac{V_{Cs}}{R_s} = C_s \frac{\Delta V_{Cs}}{T_s - T_{sn}}$$

$$\Rightarrow C_s = \frac{1}{R_s} \frac{V_{Cs}}{\Delta V_{Cs}} (T_s - T_{sn}) \approx \frac{T_s}{R_s} \frac{V_{Cs}}{\Delta V_{Cs}}. \tag{8.15}$$

In the above expression $\frac{\Delta V_{Cs}}{V_{Cs}}$ is the ratio of the snubber capacitor peak-peak ripple voltage and the snubber capacitor average voltage over a switching cycle. This value is typically chosen between 0.01 and 0.1.

The diode must be a fast-acting diode such as the Schottky diode since the conduction period T_{sn} typically lasts for a short interval.

Example 8.4

Design a snubber for the flyback converter in Example 8.1. Assume that the snubber diode is ideal and the peak input current is $I_{pk} = 2\,A$. The voltage across the transistor should not exceed 35 V under the given operating conditions.

Solution   Using Equation (8.8), under ideal conditions.

The snubber capacitor voltage must be chosen such that it is greater than the voltage across the transistor when it is off and less than the given maximum allowable voltage across the transistor, i.e.,

$$30\,V < V_{in} + V_{Cp} < 35\,V \Rightarrow 15\,V < V_{Cp} < 20\,V,$$

allowing for a snubber capacitor ripple of 10%, which is 2 V at the worst case. To keep within the specified snubber voltage limit, accounting for the ripple, the voltage is chosen to be $V_{Cp} = 18\,V$.

The snubber resistance can be obtained using Equation (8.13):

$$R_s = \frac{18(18 - 2 \times 7.5_o)}{\frac{1}{2} 2^2 \times 0.26\mu \times 100k} \approx 1\,k\Omega.$$

The power lost in the snubber circuit can be obtained using Equation (8.14):

$$P_s = \frac{V_{Cs}^2}{R_s} = \frac{18^2}{1000} = 0.324\,W,$$

which is about 6 times the leakage inductance power ($P_{Ll} = \frac{1}{2} I_{pk}^2 L_l f = 0.052\,W$), and is otherwise lost without the snubber circuit.

The snubber capacitance is obtained using Equation (8.15):

$$C_s = \frac{T_s}{R_s} \frac{V_{Cs}}{\Delta V_{Cs}} = \frac{1}{100k \times 1k} \frac{18}{0.1 \times 18} = 0.1\,\mu F.$$

8.4.3   Simulation and Hardware Prototyping—CCM with Snubber

The simulation of a nonideal flyback converter with a primary side snubber is demonstrated by means of an example:

Example 8.5

In the flyback converter shown in Figure 8.2c, $C = 490\,\mu F$, and $R = 10\,\Omega$. It is operating in DC steady state under the following conditions: $V_{in} = 15\,V$, $D = 0.5$, and $f_s = 100\,kHz$. The transformer's primary-to-secondary turns ratio is 2:1. The primary side has a magnetizing inductance of $L_m = 27.8\,\mu H$ and a leakage inductance of $L_l = 0.26\,\mu H$. The snubber resistance $R_s = 1\,k\Omega$ and the snubber capacitance $C_s = 0.1\,\mu F$. For the switch and the diode, use the parameters given in the Appendix of Chapter 2. Simulate this converter using LTspice.

Solution   The simulation file used in this example is available on the accompanying website. The LTspice model is shown in Figure 8.11, and the steady-state waveforms from the simulation of this model are shown in Figure 8.12.

The Workbench model for implementing the above example in hardware using the Sciamble lab kit is the same as the one shown in Figure 8.6. The steady-state waveforms from running the flyback converter with an RCD snubber using the Sciamble laboratory kit are shown in Figure 8.13. The step-by-step procedure for re-creating the above hardware implementation is presented in [3].

As seen in Figure 8.13, the voltage across the switch is clamped to within the 35 V design requirement given in Example 8.4.

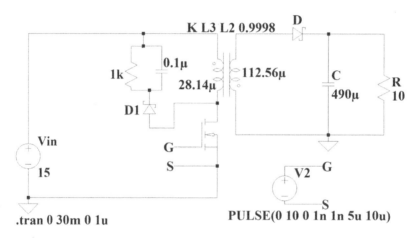

FIGURE 8.11   LTspice model.

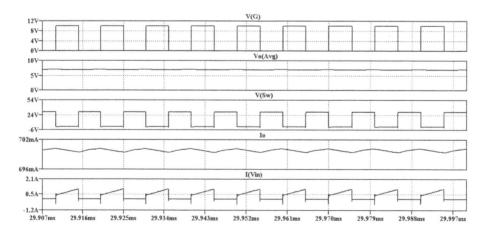

FIGURE 8.12   LTspice simulation results.

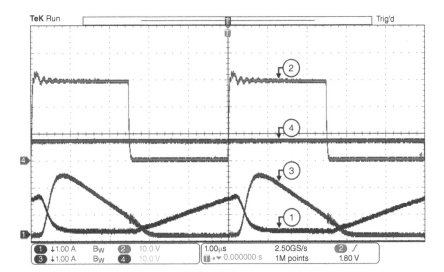

FIGURE 8.13 Workbench hardware results: (1) Input current, (2) Switch-node voltage, (3) Diode current and (4) Output voltage.

8.4.4 Simulation and Hardware Prototyping—DCM with Snubber

In the previous sections, the operation of the flyback converter was considered under continuous conduction mode. The operation of the flyback converter under discontinuous conduction mode is mostly similar to that of the buck-boost converter. The simulation of a nonideal flyback converter with a primary side snubber under discontinuous conduction mode is demonstrated by means of an example:

Example 8.6
The flyback converter in Example 8.5 is operating at a duty of $D = 0.15$ and the load resistance $R = 8\,\Omega$. Simulate this converter using LTspice.

Solution The simulation file used in this example is available on the accompanying website. The LTspice model is the same as the one shown in Figure 8.11, and the steady-state waveforms from the simulation of this model are shown in Figure 8.14.

The Workbench model for implementing the above example in hardware using the Sciamble lab kit is the same as the one shown in Figure 8.6. The steady-state waveforms from running the flyback converter with an RCD snubber using the Sciamble laboratory kit under DCM are shown in Figure 8.15.

Under DCM, when the secondary side current goes to zero, and the diode is reverse biased, the voltage across the transistor transitions from $V_{in} + nV_o$ to V_{in}. The frequency of this ringing is given by:

$$f_d = \frac{1}{2\pi\sqrt{(L_l + L_m)C_p}} \tag{8.16}$$

Many commercially available flyback converter controller ICs, such as Texas Instruments' UCC28704, typically operate under DCM to make use of the inherent ringing of the transistor voltage to turn on the transistor when the voltage across the switch is at the lowest point, thereby reducing the switching losses.

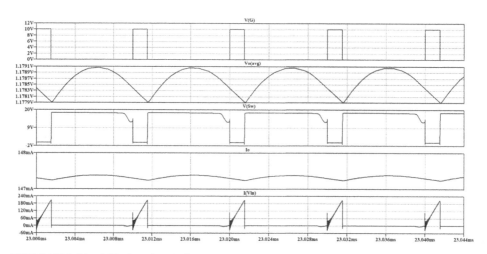

FIGURE 8.14   LTspice simulation results.

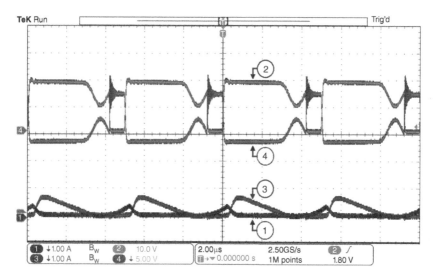

FIGURE 8.15   Workbench hardware results: (1) Input current, (2) Switch-node voltage, (3) Diode current and (4) Secondary side voltage.

So far, we have only discussed the design of snubber for the primary side. As seen in Figure 8.15, there is significant ringing in the voltage across the secondary side due to the leakage inductance and the diode's junction capacitance. The frequency of this typically tends to be much higher than that of the primary side ringing. This can be addressed by the use of a simple RC snubber across the diode, as presented in [2].

8.5   FORWARD CONVERTERS

The forward converter and its variations derived from a buck converter are commonly used in applications at low power levels up to 1 kW. A buck converter is shown in Figure 8.16a. In this circuit, a three-winding transformer is added, as shown in Figure 8.16b,

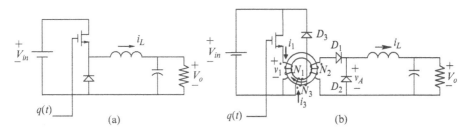

FIGURE 8.16 Buck and forward converters.

to realize a forward converter. The third winding in series with a diode D_3, and the diode D_1 are needed to demagnetize the core every switching cycle. The winding orientations in Figure 8.16b are such that the current into the dot of any of the windings will produce core flux in the same direction. We will consider steady-state converter operation in the continuous conduction mode where the output inductor current i_L flows continuously. In the following analysis, we will assume ideal semiconductor devices, $v_o(t) = V_o$, and the leakage inductances to be zero.

Initially, assuming an ideal transformer in the forward converter of Figure 8.16b, the third winding and the diode D_3 can be removed, and D_1 can be replaced by a short circuit. In such an ideal case, the forward converter operation is identical to that of the buck converter, as shown by the waveform in Figure 8.17, except for the presence of the transformer turns ratio N_2 / N_1. Therefore, in the continuous conduction mode,

$$V_o = \left(\frac{N_2}{N_1}\right) D V_{in}. \qquad (8.17)$$

In the case of a real transformer, the core must be completely demagnetized during the off interval of the transistor, hence the need for the third winding and the diodes D_1 and D_3, as shown in Figure 8.4b. Turning on the transistor causes the magnetizing flux in the core to build up, as shown in Figure 8.18. During this on-interval DT_s, D_3 gets reverse biased, thus preventing the current from flowing through the tertiary winding. The diode D_2 also gets reverse biased, and the output inductor current flows through D_1.

When the transistor is turned off, the magnetic energy stored in the transformer core forces a current to flow into the dotted terminal of the tertiary winding since the current into the dotted terminal of the secondary winding cannot flow due to D_1, which results in V_{in} being applied negatively across the tertiary winding, and the core flux to decline, as shown in Figure 8.6. (The output inductor current freewheels through D_2.) After an interval T_{demag}, the core flux comes to zero and stays zero during the remaining interval until the next cycle begins.

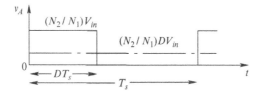

FIGURE 8.17 Forward converter operation.

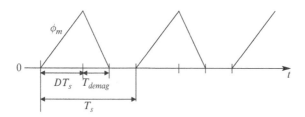

FIGURE 8.18 Forward converter core flux.

To avoid the core from saturating, T_{demag} must be less than the off interval $(1-D)T_s$ of the transistor. Typically, windings 1 and 3 are wound bifilar to provide a very tight mutual coupling between the two, and hence, $N_3 = N_1$. Therefore, a per-turn voltage of equal magnitude but opposite polarity is applied to the core during DT_s and T_{demag}. With $N_3 = N_1$, $T_{demag} = DT_s$, and at the upper limit, DT_s and T_{demag} are both equal to $T_s/2$. Therefore, with $N_3 = N_1$, the upper limit on the duty ratio is $D_{max} = 0.5$.

Example 8.7
In a forward converter, shown in Figure 8.16b, $V_{in} = 48\,\text{V}$, $V_o = 5\,\text{V}$, $N_1/N_2 = 3.5$, $N_1/N_3 = 1$, and the magnetizing inductance $L_{m1} = 150\,\mu\text{H}$. This converter is operating in equivalent CCM with a switching frequency $f_s = 200\,\text{kHz}$ and supplying an output load $P_o = 60\,\text{W}$. Assume the filter inductor current i_L to be ripple-free. Assuming this converter to be lossless, calculate the waveforms associated with it.

Solution From Equation (8.17), the duty ratio $D = 0.365$, where $T_s = 5\,\mu\text{s}$. The average currents are $I_{in} = 1.25\,\text{A}$ and $I_{out} = 12\,\text{A}$. The voltage waveforms are shown in Figure 8.19, where the output current reflected to the primary side is $(N_2/N_1)I_{out} = 3.43\,\text{A}$. The peak of the magnetizing current during the on interval DT_s can be calculated as

$$\Delta I_m = \frac{V_{in}(DT_s)}{L_{m1}} = 0.5\,\text{A}.$$

During the transistor off interval, this magnetizing current, flowing through the diode D3, decreases and comes to zero after $T_{demag} = DT_s = 1.825\,\mu\text{s}$, as shown in Figure 8.19.

8.5.1 Simulation and Hardware Prototyping

The simulation of a nonideal forward converter is demonstrated by means of an example:

Example 8.8
In the forward converter shown in Figure 8.16c, $C = 490\,\mu\text{F}$, $L = 68\,\mu\text{H}$, $R = 10\,\Omega$, $N_1/N_2 = 1$, and $N_1/N_3 = 1$. It is operating in DC steady state under the following conditions: $V_{in} = 15\,\text{V}$, $D = 0.3$, and $f_s = 100\,\text{kHz}$. The primary side has a magnetizing inductance of $L_m = 72.9\,\mu\text{H}$ and a leakage inductance of $L_l = 0.12\,\mu\text{H}$. For the switch and the diode, use the parameters given in the Appendix of Chapter 2. Simulate this converter using LTspice.

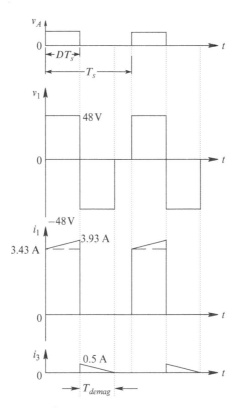

FIGURE 8.19 Waveforms in the forward converter of Example 8.7.

Solution The simulation file used in this example is available on the accompanying website. The LTspice model is, as shown in Figure 8.20, and the steady-state waveforms from the simulation of this model are shown in Figure 8.21.

The Workbench model for implementing the above example in hardware using the Sciamble lab kit is as shown in Figure 8.22.

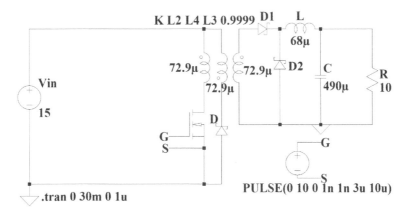

FIGURE 8.20 LTspice model.

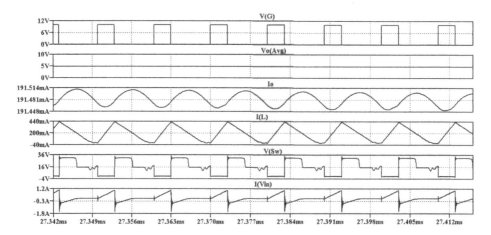

FIGURE 8.21   LTspice simulation results.

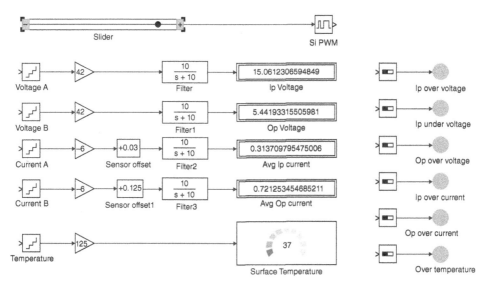

FIGURE 8.22   Workbench model.

The steady-state waveforms from running the forward converter using the Sciamble laboratory kit are shown in Figure 8.23. The step-by-step procedure for re-creating the above hardware implementation is presented in [4].

Unlike the voltage across the transistor of the flyback converter without a snubber, as seen in Figure 8.7, the voltage across the transistor of a forward converter without a snubber doesn't ring as much as seen in Figure 8.23. This is due to two reasons: the forward converter transformer typically has lower leakage inductance due to tighter coupling between the primary and the demagnetization winding, and sustained voltage greater than $2V_{in}$ across the transistor will lead to forward-biasing the demagnetization winding diode, thus effectively clamping the voltage at the sum of $2V_{in}$ and the diode forward voltage drop.

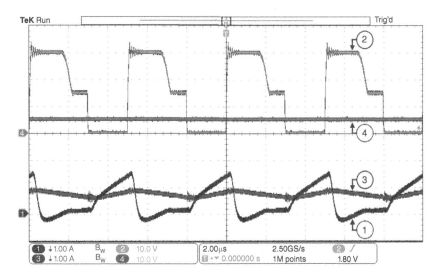

FIGURE 8.23 Workbench hardware results: (1) Input current, (2) Switch-node voltage, (3) Diode current and (4) Output voltage.

In cases where the leakage inductance of the transformer is high or the trace parasitic inductance is high due to poor layout, an RCD snubber, explained in the earlier section, can be used to limit the transient voltage spikes.

8.5.2 Two-Switch Forward Converters

Single-switch forward converters are used in power ratings up to a few hundred watts. However, two-switch forward converters discussed below eliminate the need for a separate demagnetizing winding and are used in much higher power ratings of 1 kW and even higher.

Figure 8.24 shows the topology of the two-switch forward converter, where both transistors are gated on and off simultaneously with a duty ratio $D \leq 0.5$. During the on interval DT_s, when both transistors are on, diodes D_1 and D_2 get reverse biased, and the output inductor current i_L flows through D_o, similar to that in a single-switch forward converter. During the off interval, when both transistors are turned off, the magnetizing current in the transformer core flows through the two primary-side diodes into V_{in}, thus applying V_{in} negatively to the core and causing it to demagnetize.

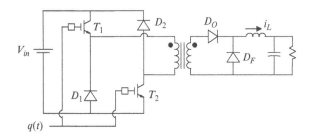

FIGURE 8.24 Two-switch forward converter.

Application of $-V_{in}$ to the primary winding causes D_o to get reverse biased, and the output inductor current i_L freewheels through D_F.

Based on the discussion regarding the demagnetization of the core in a single-switch forward converter, the switch duty ratio D is limited to 0.5. The voltage conversion ratio remains the same as in Equation (8.17).

8.6 FULL-BRIDGE CONVERTERS

Full-bridge converters consist of four transistors and hence are economically feasible only at higher power levels in applications at a few hundred watts and higher. Like forward converters, full-bridge converters are also derived from buck converters. Unlike flyback and forward converters that operate in only one quadrant of the B-H loop, full-bridge converters use the magnetic core in two quadrants.

A full-bridge converter consists of two switching power-poles, as shown in Figure 8.25, with a center-tapped transformer secondary winding. In analyzing this converter, we will assume the transformer ideal, although the effects of magnetizing current can be easily accounted for. We will consider steady-state converter operation in the continuous conduction mode where the output inductor current i_L flows continuously. As with previous converters, we will assume ideal devices and components and $v_o(t) = V_o$.

In the full-bridge converter of Figure 8.9, the voltage v_1 applied to the primary winding alternates without a DC component. The waveform of this voltage is shown in Figure 8.26, where $v_1 = V_{in}$ when transistors T_1 and T_2 are on during DT_s, and $v_1 = -V_{in}$ when T_3 and T_4 are on for an interval of the same duration. This waveform applies equal positive and negative volt-seconds to the transformer primary. The

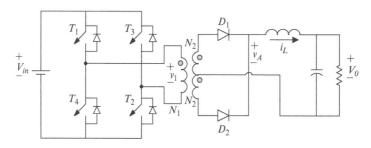

FIGURE 8.25 Full-bridge converter.

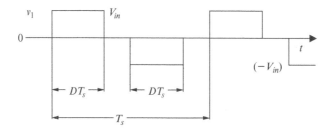

FIGURE 8.26 Full-bridge converter waveforms.

switch duty ratio D (<0.5) is controlled to achieve the output voltage regulation by means of zero intervals between the positive and the negative applied voltages.

The way the voltage across the primary winding, and hence the secondary winding, is forced to be zero classifies full-bridge converters into the following two categories:

1. Pulse-width modulated (PWM)
2. Phase-shift modulated (PSM)

PWM Control. In PWM control, all four transistors are turned off, resulting in a zero voltage across the transformer windings, as discussed shortly. With all transistors off, the output inductor current freewheels through the two secondary windings, and there are no conduction losses on the primary side of the transformer. Therefore, the PWM control results in lower conduction losses, which is why it is the control method we focus on in this chapter.

PSM Control. In phase-shift modulated control, the two transistors of each power pole are operated at nearly 50% duty ratio, with $D \approx 0.5$. The output of each power pole pulsates between V_{in} and 0 with a duty ratio of nearly 50%. The length of the zero intervals is controlled by phase-shifting the two power-pole outputs with respect to each other, as the name of this control implies. During zero intervals, either both transistors at the top or both transistors at the bottom are on, creating a short circuit (through one of the anti-parallel diodes, depending on the direction of the current) across the primary winding, resulting in $v_1 = 0$. During this short-circuited condition, the output inductor current is reflected to the primary winding and circulates through the primary-side semiconductor devices, causing additional conduction losses. However, increased conduction losses can be offset by the reduction in switching losses by this means of control, as we will discuss in detail in Chapter 10 on soft-switching.

8.6.1 PWM Control

As shown by the block diagram of Figure 8.27a, the PWM IC for full-bridge converters provides gate signals to the transistor pairs (T_1, T_2 and T_3, T_4) during alternate cycles of the ramp voltage in Figure 8.27b.

Corresponding to these PWM switching signals, the resulting sub-circuits are shown in Figure 8.28 for one-half switching cycle, where the other half-cycle is symmetric.

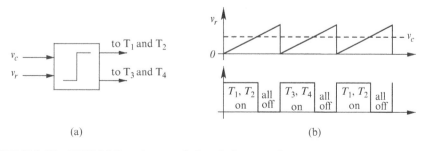

(a)   (b)

FIGURE 8.27 PWM IC and control signals for transistors.

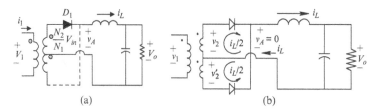

(a)   (b)

FIGURE 8.28   Full-bridge: sub-circuits.

Interval DT_s with transistors T_1, T_2 in their on state. Turning on T_1, T_2 applies a positive voltage V_{in} to the primary winding, causing D_2 to become reverse biased and i_L is carried through D_1, as shown in Figure 8.28a. During this interval, $v_A = (N_2 / N_1)V_{in}$, as plotted in Figure 8.29.

Interval $(1/2 - D)T_s$ with all transistors off. When all the transistors are turned off, there is no current in the primary winding, and the output inductor current divides equally (assuming an ideal transformer) between the two output diodes, as shown in the sub-circuit of Figure 8.28b. This ensures that the total ampere-turns acting on the transformer core equal zero because of $i_L / 2$ coming out of the dotted terminal and $i_L / 2$ going into the dotted terminal. Applying Kirchhoff's voltage law in the loop consisting of the two secondaries in Figure 8.27b shows that $v_2 + v_2' = 0$. Since $v_2 = v_2'$, the two voltages must be individually zero, and hence also the primary voltage v_1:

$$v_1 = v_2 = v_2' = 0 \tag{8.18}$$

During this interval, $v_A = 0$ as plotted in Figure 8.29.

The above discussion completes the discussion of one-half switching cycle. The other half-cycle with T_3, T_4 on applies a negative voltage $(-V_{in})$ across the primary winding for an interval DT_s and results in D_2 conducting and D_1 reverse biased. During this interval, $v_A = (N_2 / N_1)V_{in}$ as before when the positive voltage was applied to the primary winding. The waveforms during this half-cycle are as plotted in Figure 8.29.

From Figure 8.29, recognizing that $V_A = V_o$ in the DC steady state,

$$\frac{V_o}{V_{in}} = 2\left(\frac{N_2}{N_1}\right)D. \tag{8.19}$$

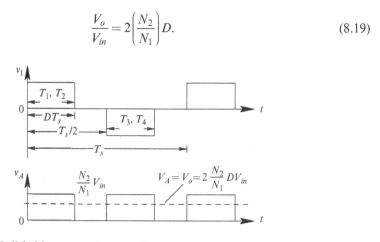

FIGURE 8.29   Full-bridge converter waveforms.

Example 8.9

In a full-bridge converter shown in Figure 8.25, $V_{in} = 48\,V$, $V_o = 5\,V$, and $N_1 / N_2 = 6$. This converter is operating in CCM with a switching frequency $f_s = 200\,kHz$ and supplying an output load $P_o = 100\,W$. The filter inductor has an inductance of $L = 0.25\,\mu H$. Assuming this converter to be lossless, calculate the waveforms associated with it.

Solution From Equation (8.19), the duty ratio $D = 0.3125$, where $T_s = 5\,\mu s$. The average currents are $I_{in} = 2.083\,A$ and $I_{out} = 20\,A$. The voltage waveforms are shown in Figure 8.14. The peak-peak ripple in the filter inductor current i_L can be calculated from the voltage waveforms in Figure 8.30,

$$\Delta I_{L,p-p} = \frac{(v_A - V_o)(DT_s)}{L} = 18.75\,A.$$

Therefore, the i_L waveform is as shown in Figure 8.30, with a minimum of $I_{L,min} = I_{out} - \frac{\Delta I_{L,p-p}}{2} = 10.625\,A$ and a maximum of $I_{L,max} = I_{out} + \frac{\Delta I_{L,p-p}}{2} = 29.375\,A$.

Taking the transformer turns ratio into account, the primary current i_1 and the input current i_{in} ramp from $1.77\,A$ to $4.896\,A$, and are zero when all the transistors are off.

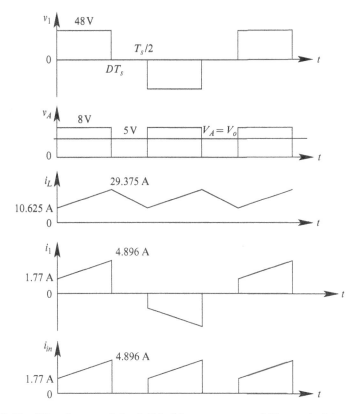

FIGURE 8.30 Waveforms of the full-bridge converter of Example 8.3.

8.6.2   Simulation and Hardware Prototyping

The simulation of a PWM full-bridge converter is demonstrated by means of an example:

Example 8.10
In the full-bridge converter shown in Figure 8.25, $C = 490\,\mu\text{F}$, $L = 68\,\mu\text{H}$, $R = 8\,\Omega$, and $N_1 / N_2 = 1$. It is operating in DC steady state under the following conditions: $V_{in} = 15\,\text{V}$, $D = 0.3$, and $f_s = 100\,\text{kHz}$. The primary side has a magnetizing inductance of $L_m = 72.9\,\mu\text{H}$ and a leakage inductance of $L_l = 0.14\,\mu\text{H}$. For the switch and the diode, use the parameters given in the Appendix of Chapter 2. Simulate this converter using LTspice.

Solution   The simulation file used in this example is available on the accompanying website. The LTspice model is shown in Figure 8.31, and the steady-state waveforms from the simulation of this model are shown in Figure 8.32.

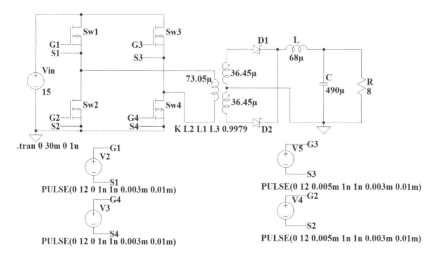

FIGURE 8.31   LTspice model.

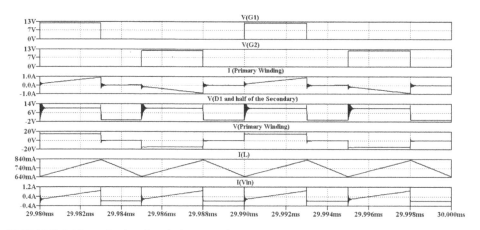

FIGURE 8.32   LTspice simulation results.

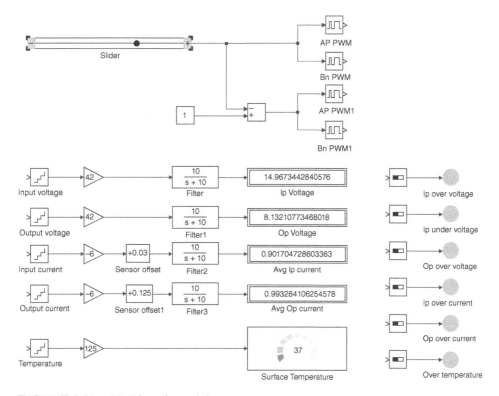

FIGURE 8.33   Workbench model.

The Workbench model for implementing the above example in hardware using the Sciamble lab kit is shown in Figure 8.33

The steady-state waveforms from running the full-bridge converter using the Sciamble laboratory kit are shown in Figure 8.34. The step-by-step procedure for re-creating the above hardware implementation is presented in [5].

8.7   HALF-BRIDGE AND PUSH-PULL CONVERTERS

Variations of full-bridge converters are shown in Figure 8.35. The half-bridge converter in Figure 8.35a consists of only two transistors but requires two split capacitors to form a DC input midpoint. It is sometimes used at slightly lower power levels compared to the full-bridge converter.

The push-pull converter in Figure 8.35b has the advantage of having both transistors' gates referenced to the low side of the input voltage. The penalty is in the transformer, where during the power transfer interval, only one-half of the primary winding and one-half of the secondary winding are utilized.

8.8   PRACTICAL CONSIDERATIONS

To provide electrical isolation between the input and the output, the feedback control loop should also have electrical isolation. There are several ways of providing this isolation, as discussed in Reference [6].

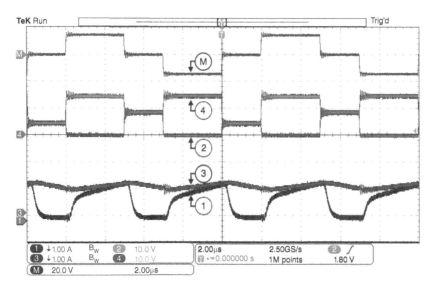

FIGURE 8.34 Workbench hardware results: (1) Input current, (2) T1, T4 switch-node voltage, (3) Secondary-side inductor current, (4) T2, T3 switch-node voltage, and (M) Transformer primary-side voltage.

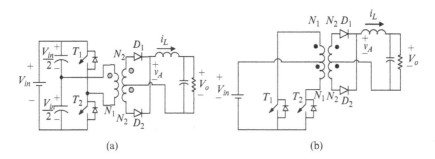

FIGURE 8.35 Half-bridge and push-pull converters.

REFERENCES

1. "Flyback Converter Lab Manual." https://sciamble.com/resources/pe-drives-lab/basic-pe/flyback-converter.
2. "Flyback Converter Snubber Design," Ray Ridley. http://www.ridleyengineering.com/images/phocadownload/12_%20flyback_snubber_design.pdf.
3. "Flyback Converter Snubber Design Lab Manual." https://sciamble.com/resources/pe-drives-lab/basic-pe/flyback-snubber-converter.
4. "Forward Converter Lab Manual." https://sciamble.com/resources/pe-drives-lab/basic-pe/forward-converter.
5. "Full-Bridge Converter Lab Manual." https://sciamble.com/resources/pe-drives-lab/basic-pe/full-bridge-converter.
6. N. Mohan, T.M. Undeland, and W.P. Robbins, *Power Electronics: Converters, Applications and Design*, 3rd Edition (New York: John Wiley & Sons, 2003).

PROBLEMS

Flyback Converters

In Problems 8.1 through 8.4, in a flyback converter, $V_{in} = 30$ V, $N_1 = 30$ turns, and $N_2 = 15$ turns. The self-inductance of winding 1 is $50\,\mu H$, and $f_s = 200$ kHz. The output voltage is regulated at $V_o = 9$ V.

8.1 Calculate and draw the waveforms shown in Figure 8.3 along with the ripple current in the output capacitor if the load is 30 W.

8.2 For the same duty ratio as in Problem 8.1, calculate the critical power, which makes this converter operate at the border of incomplete and complete demagnetization modes.

8.3 Draw the waveforms, similar to those in Problem 8.1, in Problem 8.2.

8.4 Consider that a flyback converter is operating in the complete demagnetizing mode. Under this mode of operation, derive the voltage transfer ratio V_o / V_{in} in terms of the transformer magnetizing inductance L_{m1} seen from the input-side winding, switching frequency f_s, the duty ratio D, and the load resistance R.

Forward Converter

In Problems 8.5 through 8.7, in a forward converter, $V_{in} = 30$ V, $N_1 = 10$ turns, $N_2 = 5$ turns, and $N_3 = 10$ turns. The self-inductance of winding 1 is $150\,\mu H$, and the switching frequency $f_s = 250$ kHz. The output voltage is regulated such that $V_o = 5$ V. The output filter inductance is $50\,\mu H$, and the output load is 30 W.

8.5 Calculate and draw the waveforms for v_A, i_L, i_{in}, and i_{D_3} in Figure 8.16b, where i_{in} is the current drawn from the input, and i_{D_3} is the current through the diode D_3.

8.6 If the maximum duty ratio needs to be increased to 0.7, calculate N_1 / N_3.

8.7 Why is diode D_1 necessary in Figure 8.16b?

8.8 Consider that a forward converter is operating in the discontinuous-conduction mode. Under this mode of operation, derive the voltage transfer ratio V_o / V_{in} in terms of the circuit parameters.

Two-Switch Forward Converter

In Problems 8.9 and 8.10, in a two-switch forward converter, $V_{in} = 30$ V, $N_1 / N_2 = 2$, and the switching frequency $f_s = 300$ kHz. The output voltage is regulated such that $V_o = 5$ V. The self-inductance of winding 1 is $150\,\mu H$, and the output filter inductance is $50\,\mu H$.

8.9 Calculate and draw waveforms if the output load is 200 W.

8.10 Why is the duty ratio in this converter limited to 0.5?

Full-Bridge Converter

8.11 In a full-bridge converter, shown in Figure 8.25, $V_{in} = 30$ V, $f_s = 300$ kHz, and $N_1 / N_2 = 4$. The output voltage is regulated by PWM such that $V_o = 5$ V. The output power $P_o = 250$ W, and the peak-peak ripple in the output inductor current is 10% of its average value at full load. Calculate the value of the filter inductor L. Calculate and plot all the waveforms associated with this converter. Assume the transformer ideal and the flux waveform to be symmetric with the same positive and negative peak amplitudes.

Half-Bridge Converter

8.12   A regulated DC power supply similar to Problem 8.11 is implemented using a half-bridge topology, shown in Figure 8.35a, where $N_1 / N_2 = 2$. Calculate all the waveforms associated with this converter.

Comparison of MOSFETs in Full-Bridge and Half-Bridge Converters

8.13   Compare the voltage and current ratings of MOSFETs in full-bridge and half-bridge converters of Problems 8.11 and 8.12 respectively.

Push-Pull Converter

8.14   A regulated DC power supply similar to Problem 8.11 is implemented using a push-pull topology shown in Figure 8.35b, where $N_1 / N_2 = 5$. Calculate all the waveforms associated with this converter.

Simulation Problems

8.15   Simulate a flyback converter with the following parameters and operating conditions: $V_{in} = 20$ V, $V_o = 5$ V, $N_1 / N_2 = 2$, $L_{m1} = 280\,\mu$H, the output capacitance $C = 100\,\mu$F and the switching frequency $f_s = 100$ kHz. $R_{Load} = 6\,\Omega$.
   (a)   Obtain the waveforms for i_{in}, i_{out} and v_o. What is the relationship between i_{in} and i_{out} at the time of transition from the switch to diode, and vice versa?
   (b)   What is the value of R_{crit} in this converter?
   (c)   For $R = 100\,\Omega$, obtain the waveforms for i_{in}, i_{out} and v_o.
8.16   Simulate a forward converter with the following parameters and operating conditions: $V_{in} = 20$ V, $V_o = 7.5$ V, $N_1{:}N_2{:}N_3 = 1$, $L_{m1} = 144\,\mu$H, the filter inductor $L = 100\mu$H and the output capacitance $C = 100\mu$F, and the switching frequency $f_s = 100$ kHz. $R_{Load} = 0.5\Omega$.
   (a)   Obtain the waveforms for i_1, i_3, i_{D1}, i_L, and v_o. What is the relationship between i_1, i_3, and i_{D1} at the time of transitions when the switch turns on, turns off, and the core gets demagnetized?
   (b)   What is the voltage across the switch?
   (c)   What is the value of R_{crit} in this converter?
   (d)   For $R = 50\Omega$, obtain the waveforms for i_1, i_3, i_{D1}, i_L, and v_o.
8.17   Simulate a full-bridge converter, with a midpoint rectifier, shown in Figure 8.25, with the following parameters and operating conditions: $V_{in} = 175$ V, $V_o = 4.5$ V, $N_1 / N_2 = 25$, $L_{m1} = 10$ mH, the filter inductor $L = 7.5\,\mu$H and the output capacitance $C = 100\,\mu$F, and the switching frequency $f_s = 100$ kHz. $R_{Load} = 0.5\Omega$.
   (a)   Obtain the waveforms for v_1, v_2, v_2', v_A, and v_o. What are their values when all switches are off?

(b) Plot the waveforms of i_1, i_{D1}, i_{D2}, and i_L defined in Figure 8.28. What are their values when all switches are off?

(c) Based on the converter parameters and the operating conditions, calculate the peak-peak ripple current in i_L and verify your answer with the simulation results.

DESIGN OF HIGH-FREQUENCY INDUCTORS AND TRANSFORMERS

9.1 INTRODUCTION

As discussed in Chapter 8, inductors and transformers are needed in switch-mode DC power supplies, where switching frequencies are in excess of 100 kHz. High-frequency inductors and transformers are generally not available off-the-shelf and must be designed based on the application specifications. A detailed design discussion is presented in [1]. In this chapter, a simple and commonly used approach called the area-product method is presented, where the thermal considerations are ignored. This implies that the magnetic component built on the design basis presented here should be evaluated for its temperature rise and efficiency, and the core and the conductor sizes should be adjusted accordingly.

9.2 BASICS OF MAGNETIC DESIGN

In designing high-frequency inductors and transformers, a designer is faced with countless choices. These include the choice of core materials, core shapes (some offer better thermal conduction whereas others offer better shielding to stray flux), cooling methods (natural convection versus forced cooling), and losses (lower losses offer higher efficiency at the expense of larger size and higher weight), to name a few. However, all magnetic design-optimization programs calculate two basic quantities from given electrical specifications:

1. The peak flux density B_{max} in the magnetic core to limit core losses, and
2. The peak current density J_{max} in the winding conductors to limit conduction losses.

Power Electronics A First Course: Simulations and Laboratory Implementations, Second Edition.
Ned Mohan and Siddharth Raju.
© 2023 John Wiley & Sons, Inc. Published 2023 by John Wiley & Sons, Inc.
Companion Website: www.wiley.com/go/mohan/powerelectronics2e

The design procedure presented in this chapter assumes values for these two quantities based on the intended applications of inductors and transformers. However, they may be far from optimal in certain situations.

9.3 INDUCTOR AND TRANSFORMER CONSTRUCTION

Figures 9.1a and 9.1b represent the cross-section of an inductor and a transformer wound on toroidal cores. In Figure 9.1a, for an inductor, the same current i passes through all N turns of a winding. In the transformer of Figure 9.1b, there are two windings where the current i_1 in winding 1, with N_1 bigger cross-section conductors, is in the opposite direction to that of i_2 in winding 2 with N_2 smaller cross-section conductors. In each winding, the conductor cross-section is chosen such that the peak current density J_{max} is not exceeded at the maximum specified current in that winding. The core area A_{core} in Figures 9.1a and 9.1b allows the flow of flux lines without exceeding the maximum flux density B_{max} in the core.

9.4 AREA-PRODUCT METHOD

The area-product method, based on preselected values of the peak flux density B_{max} in the core and the peak current density J_{max} in the conductors, allows an appropriate core size to be chosen, as described below.

9.4.1 Core Window Area A_{window}

The windows of the toroidal cores in Figures 9.1a and 9.1b accommodate the winding conductors, where the conductor cross-sectional area A_{cond} depends on the maximal RMS current for which the winding is designed. In the expression for the window area below, the window fill factor k_w in a range from 0.3 to 0.6 accounts for the fact that the entire area of the window cannot be filled, and the subscript y designates a winding, where in general, there may be more than one, as in a transformer:

$$A_{window} = \frac{1}{k_w} \sum_y \left(N_y \, A_{cond,y} \right). \tag{9.1}$$

In Equation (9.1), the conductor cross-sectional area in winding y depends on its maximal RMS current and the maximal allowed current density J_{max} that is generally chosen to be the same for all windings:

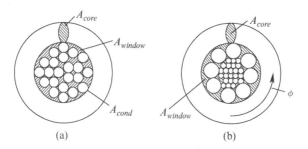

(a)   (b)

FIGURE 9.1 Cross-sections.

$$A_{cond,y} = \frac{I_{rms,y}}{J_{max}}.$$ (9.2)

Substituting Equation (9.2) into Equation (9.1),

$$A_{window} = \frac{\sum_y (N_y I_{rms,y})}{k_w J_{max}},$$ (9.3)

which shows that the window area is linearly proportional to the number of turns chosen by the designer.

9.4.2 Core Cross-Sectional Area A_{core}

The core cross-sectional area in Figures 9.1a and 9.1b depends on the peak flux $\hat{\phi}$ and the choice of the maximal allowed flux density B_{max} to limit core losses:

$$A_{core} = \frac{\hat{\phi}}{B_{max}}.$$ (9.4)

How the flux is produced depends on whether the device is an inductor or a transformer. In an inductor, $\hat{\phi}$ depends on the peak current and equals the peak flux linkage $N\hat{\phi}$. Hence,

$$\hat{\phi} = \frac{L\hat{I}}{N} \quad \text{(inductor)}.$$ (9.5)

In a transformer, based on Faraday's law, the flux depends linearly on the applied volt-seconds and inversely on the number of turns. This is shown in Figure 9.2 for a forward converter transformer with $N_1 = N_3$, and the duty ratio D, which is limited to 0.5. Therefore, we can express the peak flux in Figure 9.2 as

$$\hat{\phi} = \frac{k_{conv} V_{in}}{N_1 f_s} \quad \text{(transformer)},$$ (9.6)

where the factor k_{conv} equals D in a forward converter and typically has a maximum value of 0.5. The factor k_{conv} can be derived for transformers in other converter

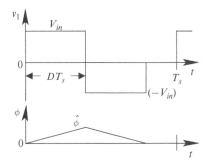

FIGURE 9.2 Waveforms in a transformer for a forward converter.

topologies based on the specified operating conditions, for example, it equals $D/2$ in a full-bridge converter. In general, the peak flux can be expressed in terms of any one of the windings, y, for example, as

$$\hat{\phi} = \frac{k_{conv} V_y}{N_y f_s} \quad \text{(transformer)}. \tag{9.7}$$

Substituting for $\hat{\phi}$ from Equations (9.5) and (9.7) into Equation (9.4), respectively, we find:

$$A_{core} = \frac{L\hat{I}}{NB_{max}} \quad \text{(inductor)} \tag{9.8}$$

$$A_{core} = \frac{k_{conv} V_y}{N_y f_s B_{max}} \quad \text{(transformer)} \tag{9.9}$$

Equations (9.8) and (9.9) show that in both cases, the core cross-sectional area is inversely proportional to the number of turns chosen by the designer.

9.4.3   Core Area-Product $A_P (= A_{core} A_{window})$

The core area-product is obtained by multiplying the core cross-sectional area A_{core} with its window area A_{window}:

$$A_p = A_{core} A_{window}. \tag{9.10}$$

Substituting for A_{window} and A_{core} from the previous equations,

$$A_p = \frac{L\hat{I} I_{rms}}{k_w J_{max} B_{max}} \quad \text{(inductor)} \tag{9.11}$$

$$A_p = \frac{k_{conv} \sum_y (V_y I_{y,rms})}{k_w B_{max} J_{max} f_s} \quad \text{(transformer)}. \tag{9.12}$$

Equations (9.11) and (9.12) show that the area-product that represents the overall size of the device is independent (as it ought to be) of the number of turns. After all, the core and the overall component size should depend on the electrical specifications and the assumed values of B_{max} and J_{max} and *not* on the number of turns, which is an internal design variable.

9.4.4   Design Procedure Based on Area-Product A_p

Once we pick the appropriate material and the shape for a core, the cores by various manufacturers are cataloged based on the area-product A_p. Having calculated the value of A_p above, we can select the appropriate core. It should be noted that there are infinite combinations of the core cross-sectional area A_{core} and the window area A_{window} that yield the desired area-product A_p. However, manufacturers take pains in producing cores such that for a given A_p, a core has A_{core} and A_{window} that are individually optimized for power density. Once we select a core, it has specific A_{core} and A_{window}, which allow the number of turns to be calculated as follows:

$$N = \frac{L\hat{I}}{B_{\max} A_{core}} \quad \text{(inductor; from Equation 9.8)} \tag{9.13}$$

$$N_y = \frac{k_{conv}V_y}{A_{core} f_s B_{\max}} \quad \text{(inductor; from Equation 9.9).} \tag{9.14}$$

In an inductor, to ensure that it has the specified inductance, an air gap of an appropriate length ℓ_g is introduced in the path of flux lines. Assuming the chosen core material to have very high permeability, the core inductance is primarily dictated by the reluctance $\Re_g$ of the air gap, such that

$$L \simeq \frac{N^2}{\Re_g}, \tag{9.15}$$

where

$$\Re_g \simeq \frac{\ell_g}{\mu_o A_{core}}. \tag{9.16}$$

Using Equations (9.15) and (9.16), the air gap length ℓ_g can be calculated as:

$$\ell_g = \frac{N^2 \mu_o A_{core}}{L}. \tag{9.17}$$

The above equations are approximate because they ignore the effects of finite core permeability and the fringing flux, which can be substantial. Core manufacturers generally specify measured inductance as a function of the number of turns for various values of the air gap length. In this section, we used a toroidal core for descriptive purposes in which it will be difficult to introduce an air gap. If a toroidal core must be used, it can be picked with a distributed air gap such that it has the effective air gap length as calculated above. The above procedure explained for toroidal cores is equally valid for other types of cores. The actual design described in the next section illustrates the introduction of an air gap in a pot core.

9.5 DESIGN EXAMPLE OF AN INDUCTOR

In this example, we will discuss the design of an inductor that has an inductance $L = 100\,\mu\text{H}$. The worst-case current through the inductor is shown in Figure 9.3, where the average current $I = 5.0\,\text{A}$, and the peak-peak ripple $\Delta I = 0.75\,\text{A}$ at the switching

FIGURE 9.3 Inductor current waveforms.

frequency $f_s = 100\,\text{kHz}$. We will assume the following maximum values for the flux density and the current density: $B_{max} = 0.25\,\text{T}$, and $J_{max} = 6.0\,\text{A}\,/\,\text{mm}^2$ (for larger cores, this is typically in a range of 3 to $4\,\text{A}\,/\,\text{mm}^2$). The window fill factor is assumed to be $k_w = 0.5$.

The peak value of the inductor current from Figure 9.3 is $\hat{I} = I + \dfrac{\Delta I}{2} = 5.375\,\text{A}$.

The RMS value of the current for the waveform shown in Figure 9.3 can be calculated as $I_{rms} = \sqrt{I^2 + \dfrac{1}{12}\Delta I^2} \simeq 5.0\,\text{A}$ (the derivation is left as a homework problem).

From Equation (9.11),

$$\text{Area - Product} \quad A_p = \frac{100 \times 10^{-6} \times 5.375 \times 5}{0.5 \times 0.25 \times 6 \times 10^6} \times 10^{12} = 3587\,\text{mm}^4$$

From the Magnetics, Inc. catalog [2], we will select a p-type material that has a saturation flux density of $0.5\,\text{T}$ and is quite suitable for use at the switching frequency of $100\,\text{kHz}$. A pot core 26×16, which is shown in Figure 9.4 for a laboratory experiment, has the core Area $A_{core} = 93.1\,\text{mm}^2$ and the window Area $A_{window} = 39\,\text{mm}^2$. Therefore, we will select this core, which has an Area-Product $A_p = 93.1 \times 39 = 3631\,\text{mm}^4$. From Equation (9.13),

$$N = \frac{100\mu \times 5.375}{0.25 \times 93.1 \times 10^{-6}} \simeq 23\,\text{Turns}.$$

Winding wire cross-sectional area $A_{cond} = I_{rms}\,/\,J_{max} = 5.0\,/\,6.0 = 0.83\,\text{mm}^2$. We will use five strands of American Wire Gauge AWG 25 wires [3], each with a

FIGURE 9.4 Pot core mounted on a plug-in board.

cross-sectional area of 0.16 mm^2, in parallel. From Equation (9.17), the air gap length can be calculated as

$$\ell_g = \frac{23^2 \times 4\pi \times 10^{-7} \times 93.1 \times 10^{-6}}{100\mu} \simeq 0.62\,\text{mm}.$$

9.6 DESIGN EXAMPLE OF A TRANSFORMER FOR A FORWARD CONVERTER

The required electrical specifications for the transformer in a forward converter are as follows: $f_s = 100\,\text{kHz}$ and $V_1 = V_2 = V_3 = 30\,\text{V}$. Assume the RMS value of the current in each winding to be 2.5 A. We will choose the following values for this design: $B_{\max} = 0.25\,\text{T}$ and $J_{\max} = 5\,\text{A/mm}^2$. From Equation (9.12), where $k_w = 0.5$ and $k_{conv} = 0.5$,

$$A_p = \frac{k_{conv}}{k_w \, f_s B_{\max} J_{\max}} \sum_y \hat{V}_y \, I_{\text{rms},y} = 1800\,\text{mm}^4.$$

For the pot core 22 × 13 [2], $A_{core} = 63.9\,\text{mm}^2$, $A_{window} = 29.2\,\text{mm}^2$, and therefore $A_p = 1866\,\text{mm}^4$. For this core, the winding wire cross-sectional area is obtained as

$$A_{cond,1} = \frac{I_{1,\text{rms}}}{J_{\max}} = \frac{2.5}{5} = 0.5\,\text{mm}^2.$$

We will use three strands of AWG 25 wires [3], each with a cross-sectional area of 0.16 mm^2, in parallel for each winding. From Equation (9.14),

$$N_1 = \frac{0.5 \times 30}{\left(63.9 \times 10^{-6}\right) \times \left(100 \times 10^3\right) \times 0.25} \simeq 10.$$

Hence,

$$N_1 = N_2 = N_3 = 10$$

9.7 THERMAL CONSIDERATIONS

Designs presented here do not include eddy current losses in the windings, which can be very substantial due to proximity effects. These proximity losses in a conductor are due to the high-frequency magnetic field generated by other conductors in close proximity. These proximity losses can be minimized by designing inductors with a single-layer construction. In transformers, windings can be interleaved to minimize these losses, as described in detail in [1]. Therefore, the area-product method discussed in this chapter is a good starting point, but the designs must be evaluated for the temperature rise due to additional losses. A more detailed analysis is presented in [4].

REFERENCES

1. N. Mohan, T.M. Undeland, and W.P. Robbins, *Power Electronics: Converters, Applications and Design*, 3rd Edition (New York: John Wiley & Sons, 2003).
2. Magnetics, Inc. Ferrite Cores: www.mag-inc.com.
3. Wire Gauge Comparison Chart: http://www.engineeringtoolbox.com/wire-gauges-d_419. html.
4. N. Mohan, W. Robins, T. Undeland, and S. Raju, *Power Electronics for Grid-Integration of Renewables: Analysis, Simulations and Hardware Lab* (New York: John Wiley & Sons, 2023).

PROBLEMS

Inductor Design

9.1   Derive the expression for the RMS current for the current waveform in Figure 9.3.

9.2   In the design example of the inductor in Section 9.5 of this chapter, the core has an area $A_{core} = 93.1 \, \text{mm}^2$, the mean magnetic path length $\ell = 37.6 \, \text{mm}$, and the relative permeability of the core material is $\mu_r = \mu_m / \mu_o = 5000$. Calculate the inductance with 23 turns if the air gap is not introduced in this core in the flux path.

9.3   In the inductor design presented in Section 9.5 of this chapter, what is the reluctance offered by the magnetic core as compared to that offered by the air gap if the mean magnetic path length of the core is $\ell = 37.6 \, \text{mm}$?

9.4   In Problem 9.2, what is the maximum current that will cause the peak flux density to reach 0.25 T?

9.5   In the inductor designed in Section 9.5 of this chapter, what will be the inductance and the maximum current that can be passed without exceeding the B_{max} specified, if the air gap introduced by mistake is only one-half of the required value?

Transformer Design

9.6   In the design example of the transformer in Section 9.6 of this chapter, the core has an area $A_{core} = 63.9 \, \text{mm}^2$, the magnetic path $\ell = 31.2 \, \text{mm}$, and the relative permeability of the core material is $\mu_r = \mu_m / \mu_o = 5000$. Calculate the peak magnetizing current at duty ratio of 0.5 in this example.

9.7   What is the tertiary winding conductor diameter needed for the magnetizing current calculated in Problem 9.6?

9.8   Derive k_{conv} for transformers in two-switch forward, half-bridge, and push-pull converters.

10

SOFT-SWITCHING IN DC-DC CONVERTERS AND HALF-BRIDGE RESONANT CONVERTERS

10.1 INTRODUCTION

In converters so far, we have discussed hard-switching in the switching power-pole, as described in Chapter 2. In this chapter, we will look at the problems associated with hard-switching and some of the practical circuits where this problem can be minimized with soft-switching.

10.2 HARD-SWITCHING IN SWITCHING POWER POLES

Figure 10.1a shows the switching power-pole, and Figure 10.1b shows the hard-switching waveforms, which were discussed in Chapter 2. Because of the simultaneous high voltage and high current associated with the transistor during the switching transition, the switching power losses in the transistor increase linearly proportional to the switching frequency and the sum of the times $t_{c(on)}$ and $t_{c(off)}$, shown in Figure 10.1b, assuming an ideal diode,

$$P_{sw} \propto f_s \big(t_{c(on)} + t_{c(off)} \big). \tag{10.1}$$

In hard-switching converters, in addition to the switching power losses decreasing the energy efficiency, the other problems are device stresses, thermal management of power losses, and electromagnetic interference resulting from high di/dt and dv/dt due

Power Electronics A First Course: Simulations and Laboratory Implementations, Second Edition.
Ned Mohan and Siddharth Raju.
© 2023 John Wiley & Sons, Inc. Published 2023 by John Wiley & Sons, Inc.
Companion Website: www.wiley.com/go/mohan/powerelectronics2e

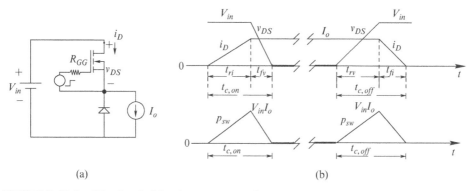

(a)   (b)

FIGURE 10.1   Hard switching in a power pole.

to fast transitions in the converter voltages and currents. The above problems are exacerbated by the stray capacitances and leakage inductances associated with the converter layout and the components.

In order to reduce the overall converter size while maintaining high energy efficiency, the trend is to design DC-DC converters operating at as high a switching frequency as possible (typically 100–200 kHz in small power ratings), using fast-switching MOSFETs. At high switching frequency, the switching power losses become unacceptable if hard-switching is used, and hence soft-switching is often employed, as briefly described in this chapter.

The problems described above associated with hard-switching can be minimized by means of the following:

- Circuit layout to reduce stray capacitances and inductances.
- Snubbers to reduce di/dt and dv/dt.
- Gate-drive control to reduce di/dt and dv/dt.
- Soft-switching.

It is always recommended to have a layout to reduce stray capacitances and inductances. Snubbers consist of passive elements (R, C, and possibly a diode) to reduce di/dt and dv/dt during the switching transient, by shaping the switching trajectory. The trend in modern power electronics is to use snubbers only in transformer-isolated DC-DC converters, where the leakage inductance associated with the high-frequency transformer can be substantial, in spite of a good circuit layout. Generally, snubbers do not reduce the overall losses; rather, they shift some of the switching losses in the transistor to the snubber resistor. A detailed discussion of snubbers is beyond the scope of this book; the interested reader is advised to refer to [1].

By controlling the gate voltage of MOSFETs and IGBTs, it is possible to slow down the turn-on and turn-off speed, thereby resulting in reduced di/dt and dv/dt, at the expense of higher switching losses in the transistor. The above techniques, at best, result in a partial solution to the problems of hard-switching.

However, there are certain topologies and control, as described in the next section, that allow soft-switching that essentially eliminates the drawbacks of hard-switching without creating new problems.

10.3   SOFT-SWITCHING IN SWITCHING POWER-POLES

There are many such circuits and control techniques proposed in the literature, most of which may make the problem of EMI and the overall losses worse due to large conduction losses in the switches and other passive components. Avoiding these topologies, only a few soft-switching circuits are practical.

The goal in soft-switching is that the switching transition in the power pole occurs under very favorable conditions, that is, the switching transistor has a zero voltage and/or zero current associated with it. Based on these conditions, the soft-switching circuits can be classified as follows:

- ZVS (zero-voltage switching), and
- ZCS (zero-current switching)

We will consider only the converter circuits using MOSFETs, which result in ZVS (zero-voltage switching). The reason is that, based on Equation 10.1, soft-switching is of interest at high switching frequencies where MOSFETs are used. In MOSFETs operating at high switching frequencies, a significant reduction of switching loss is achieved by *not* dissipating the charge associated with the junction capacitance inside the MOSFET each time it turns on. This implies that for meaningful soft-switching, the MOSFETs should be turned on under a zero-voltage switching (ZVS) condition. As we will see shortly, the turn-off also occurs at ZVS.

10.3.1   Zero-Voltage Switching (ZVS)

To illustrate the ZVS principle, the intrinsic anti-parallel diode of the MOSFET is shown as being distinct in Figures 10.2a and 10.2b, where a capacitor is used in parallel. This MOSFET is connected in a circuit such that before applying the gate voltage to turn the MOSFET on, the switch voltage is brought to zero, and the anti-parallel diode is conducting, as shown in Figure 10.2a. This results in an ideal lossless turn-on at ZVS. At turn-off, as shown in Figure 10.2b, the capacitor across the switch results in an essentially ZVS turn-off where the current through the MOSFET channel is removed while the voltage across the device remains small (essentially zero) due to the parallel capacitor. This ZVS principle is illustrated by modifying the synchronous-rectified buck DC-DC converter, as discussed below.

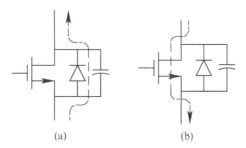

(a)   (b)

FIGURE 10.2   ZVS in a MOSFET.

10.3.2 Synchronous Buck Converter with ZVS

As discussed in section 3.10 (Chapter 3), synchronous-rectified buck DC-DC converters are used in applications where the output voltage is very low. Figure 10.3a shows a synchronous-rectified DC-DC converter where the diode has been replaced by another MOSFET, and the two MOSFETs are provided complementary gate signals q^+ and q^-. The waveforms associated with this synchronous buck converter are shown in Figure 10.3b, where the inductor is large such that the inductor-current ripple is small (shown by the solid curve in Figure 10.3b), and the inductor current remains positive in the direction shown in Figure 10.3a in the continuous conduction mode.

To achieve ZVS, the circuit of Figure 10.3a is modified, as shown in Figure 10.4a, by showing the internal diode of the MOSFET explicitly and adding small external capacitances (in addition to the junction capacitances inherent in MOSFETs). The inductance value in this circuit is chosen to be much smaller such that the inductor current has a waveform shown dotted in Figure 10.3b with a large ripple and such that the current i_L is both positive and negative during every switching cycle.

We will consider the transition at time $t = 0$ labeled in Figure 10.3b and Figure 10.4b, when the inductor current is at its peak $\hat{I}_L$ in Figure 10.4a. The gate signal q^+ of the transistor T^+, which is initially conducting $\hat{I}_L$, goes to zero, while q^- remains zero, as shown in Figure 10.4b. The transition time during which the current transfers from T^+ to T^- is very short, and it is reasonable to assume for discussion purposes that the inductor current remains constant at $\hat{I}_L$, as shown in Figure 10.5a.

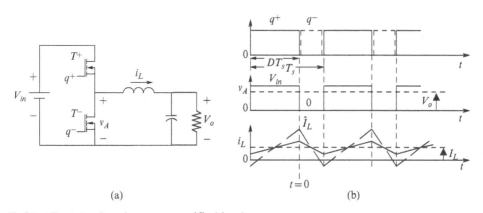

(a)   (b)

FIGURE 10.3 Synchronous-rectified buck converter.

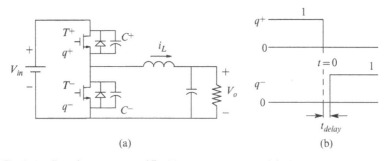

(a)   (b)

FIGURE 10.4 Synchronous-rectified buck converter with ZVS.

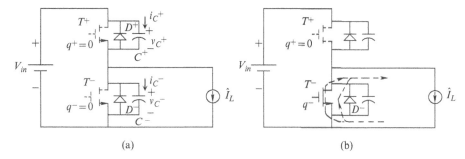

(a)   (b)

FIGURE 10.5 Transition in synchronous-rectified buck converter with ZVS.

In the circuit of Figure 10.5a with both q^+ and q^- equal to zero, initially $v_{C^+}(0) = 0$ and $v_{C^-}(0) = V_{in}$, where from Kirchhoff's law, these two voltages must add to the input voltage,

$$v_{C^+} + v_{C^-} = V_{in}. \tag{10.2}$$

As the current through T^+ declines, equal and opposite currents flow through the two capacitors in Figure 10.5a, which can be derived from Equation 10.2 as follows, assuming equal capacitances C: differentiating both sides of Equation 10.2 and multiplying both sides by C,

$$C\frac{d}{dt}v_{C^+} + C\frac{d}{dt}v_{C^-} = 0 \tag{10.3}$$

$$i_{C^+} + i_{C^-} = 0 \quad \Rightarrow \quad i_{C^+} = -i_{C^-}. \tag{10.4}$$

As the current through T^+ declines, a positive i_{C^+} causes v_{C^+} to rise from 0, and a negative i_{C^-} causes v_{C^-} to decline from its initial value of V_{in}. If this voltage transition happens slowly compared to the current fall time of the MOSFET T^+, then the turn-off of T^+ is achieved at essentially zero voltage (ZVS). After the current through T^+ has gone to zero, applying Kirchhoff's current law in Figure 10.5a and using Equation 10.4 results in

$$i_{C^+} = -i_{C^-} = \frac{\hat{I}_L}{2}. \tag{10.5}$$

In Figure 10.5a, during the turn-off transition of T^+, v_{C^+} rises to V_{in}, and v_{C^-} declines to 0. The voltage v_{C^-} cannot become negative because of the diode D^- (assuming an ideal diode with zero forward voltage drop), which begins to conduct the entire i_L in Figure 10.5b, marking the ZVS turn-off of T^+.

Once D^- begins to conduct, the voltage is zero across T^-, which is applied a gate signal q^- to turn on, as shown in Figure 10.4b, thus resulting in the ZVS turn-on of T^-. Subsequently, the entire inductor current begins to flow through the channel of T^- in Figure 10.5b. The important item to note here is that the gate signal to T^- is appropriately delayed by an interval T_{delay}, shown in Figure 10.4b, making sure that q^- is applied after D^- begins to conduct.

The next half-cycle in this converter is similar to the ZVS turn-off of T^-, followed by the ZVS turn-on of T^+, facilitated by the negative peak of the inductor current.

Although this buck converter results in ZVS turn-on and turn-off of both transistors, the inductor current has a large ripple, which will also make the size of the filter capacitor large since it has to carry the inductor current ripple. In order to make this circuit practical, the overall ripple that the output capacitor has to carry can be made much smaller (similar ripple reduction occurs in the current drawn from the input source) by an interleaving of two or more such converters, as discussed in section 3.11 (Chapter 3). Another advantage of this converter is its ability to respond rapidly to load change.

10.3.3 Phase-Shift Modulated (PSM) DC-DC Converters

Another practical soft-switching technology is the phase-shift modulated (PSM) DC-DC converter, shown in Figure 10.6a. It is a variation of the PWM DC-DC converters discussed in Chapter 8.

The switches in each power pole of the PSM converter operate at nearly 50% duty ratio, and the regulation of the output voltage is provided by shifting the output of one switching power-pole with respect to the other, as shown in Figure 10.6b, to control the zero-voltage intervals in the transformer primary voltage v_{AB}.

To provide the ZVS turn-on and turn-off of switches, a capacitor is placed across each switch. This topology makes use of the transformer leakage inductance and the magnetizing current.

10.4 HALF-BRIDGE RESONANT CONVERTER

It is possible to achieve soft-switching in converters for induction heating and compact fluorescent lamps by using a half-bridge converter with soft-switching, as shown in Figure 10.7.

This converter consists of two DC capacitors that establish the midpoint such that each capacitor has $V_{in}/2$ across them. The load in this generic circuit is represented by a resistance R in series with the resonant circuit elements L and C, which have a resonant frequency

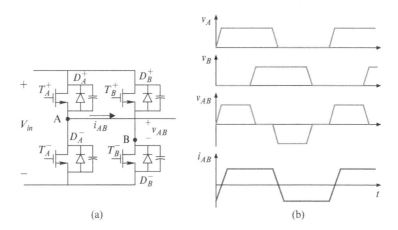

(a)   (b)

FIGURE 10.6 Phase-shift modulated (PSM) DC-DC converter.

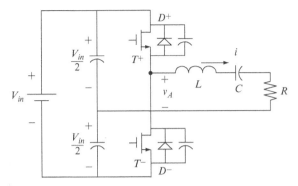

FIGURE 10.7   Half-bridge resonant converter.

$$f_o = \frac{1}{2\pi} \frac{1}{\sqrt{LC}}. \tag{10.6}$$

The switching frequency f_s is selected to be greater than the resonant frequency:

$$f_s > f_o. \tag{10.7}$$

In Figure 10.7, two switches are switched with essentially 50% duty ratio. This results in v_A that has a square waveform, as shown in Figure 10.8. Because the switching frequency in the resonant circuit is higher than the resonant frequency, this circuit appears inductive at the switching frequency. Therefore, the current i lags behind the voltage waveform, as shown in Figure 10.8. Since the impedance of the resonant circuit will be much higher at the harmonic frequency of the input voltage, the current will essentially have a sinusoidal waveform, as shown in Figure 10.8.

At $t = 0$, T^- is forced to turn off, essentially at ZVS due to lossless capacitors, and after a brief period during which capacitor voltages change, the current begins to flow through D^+. As soon as D^+ begins to conduct, T^+ is gated on, and the current can flow through it when it becomes positive. At $T_s/2$, T^+ is forced to turn off at essentially ZVS, and the next half-cycle ensues.

A detailed description of such resonant circuits is provided in [1, 2].

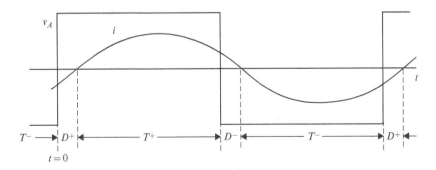

FIGURE 10.8   Converter output waveforms.

REFERENCES

1. N. Mohan, T.M. Undeland, and W.P. Robbins, *Power Electronics: Converters, Applications and Design*, 3rd Edition (New York: John Wiley & Sons, 2003).
2. N. Mohan, W. Robins, T. Undeland, and S. Raju, *Power Electronics for Grid-Integration of Renewables: Analysis, Simulations and Hardware Lab* (New York: John Wiley & Sons, 2023).

PROBLEMS

10.1   In a synchronous-rectified buck converter with ZVS, shown in Figure 10.4a, $V_{in} = 12\,V$, $V_o = 5\,V$, $f_s = 100\,kHz$, and the maximum load is $25\,W$. Calculate the filter inductance such that the negative peak current through the inductor is at least 1.5 amps.

10.2   In Problem 10.1, calculate the capacitances across the MOSFETs if the charge/discharge time is to be no more than $0.5\,\mu s$.

SIMULATION PROBLEMS

10.3   In a synchronous buck converter, such as that shown in Figure 10.4a, $V_{in} = 21\,V$, $V_o = 10\,V$, the filter-inductor $L = 20\,\mu H$ and the output capacitance $C = 1,000\,\mu F$, and the switching frequency $f_s = 100\,kHz$. $R_{Load} = 10.0\,\Omega$. The gate voltages are as shown below:

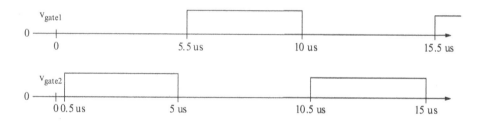

(a)   Obtain the waveforms for v_A and i_L.
(b)   Obtain the voltage across and the current through one of the switches. Comment on the zero-voltage/current switchings.
(c)   Around the blanking time, obtain the currents through one of the switches and through its associated diode and the snubber capacitors.

10.4   In a resonant converter, shown in Figure 10.7, $V_{in} = 155\,V$, $L = 100\,\mu H$, $C = 17.4\,nF$ and $R = 15.0\,\Omega$. The snubber capacitors across the MOSFETs are $1.0\,nF$ each. The switching frequency $f_s = 125\,kHz$.
(a)   Obtain the waveforms for v_A and i_L.
(b)   By Fourier analysis, obtain and plot v_{A1} and i_{L1} waveforms, where the subscript 1 refers to the fundamental frequency component. Why does the current lag the voltage?
(c)   Plot the voltage across the capacitor and the current through it.

11

APPLICATIONS OF SWITCH-MODE POWER ELECTRONICS IN MOTOR DRIVES, UNINTERRUPTIBLE POWER SUPPLIES, AND POWER SYSTEMS

11.1 INTRODUCTION

Electric motor drives (AC and DC), uninterruptible power supplies (UPS), and power systems are major application areas of switch-mode power electronics. In these applications, the role of power electronics converters is to synthesize appropriate voltages of desirable amplitude, frequency, phase, and the number of phases, where the power flow is often bidirectional. This synthesis process is described in detail in the next chapter. The purpose of this chapter is to present various applications and describe the voltage and current requirements they impose on the power electronics converters.

11.2 ELECTRIC MOTOR DRIVES

As described in Chapter 1, motor drives represent an important application area of power electronics with a market value of tens of billions of dollars annually. Figure 11.1 shows the block diagram of an electric motor drive, or for short, an electric drive. In response to an input command, electric drives efficiently control the speed and the

Power Electronics A First Course: Simulations and Laboratory Implementations, Second Edition.
Ned Mohan and Siddharth Raju.
© 2023 John Wiley & Sons, Inc. Published 2023 by John Wiley & Sons, Inc.
Companion Website: www.wiley.com/go/mohan/powerelectronics2e

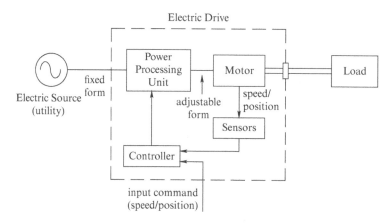

FIGURE 11.1   Block diagram of an electric drive system [1].

position of the mechanical load. The controller, by comparing the input command for speed and position with the actual values measured through sensors, provides appropriate control signals to the power-processing unit (PPU) consisting of power semiconductor devices.

As Figure 11.1 shows, the power-processing unit gets its power from the utility source with single-phase or three-phase sinusoidal voltages of a fixed frequency and constant amplitude. The power-processing unit, in response to the control inputs, efficiently converts these fixed-form input voltages into an output of the appropriate form (in frequency, amplitude, and number of phases) that is optimally suited for operating the motor. The input command to the electric drive in Figure 11.1 may come from a process computer, which considers the objectives of the overall process and issues a command to control the mechanical load. However, in general-purpose applications, electric drives operate in an open-loop manner without any feedback.

We will briefly examine various types of electric machines in terms of their terminal characteristics in steady state in order to determine the voltage and current ratings in designing the power electronics interface.

11.2.1 DC Motors

DC motors were widely used in the past for all types of drive applications, and they continue to be used for controlling speed and position. There are two designs of DC machines: stators consisting of either permanent magnets or a field winding. The power-processing units can also be classified into two categories, either switch-mode power converters that operate at a high switching frequency, as discussed in the next chapter, or line-commutated thyristor converters, which are discussed later, in Chapter 14. In this chapter, we will describe permanent-magnet DC motors, which are usually supplied by switch-mode power electronic converters.

Figure 11.2 shows a cut-away view of a DC motor. It shows a permanent-magnet stator, a rotor that carries a winding, a commutator, and the brushes. In DC machines, the stator establishes a uniform flux ϕ_f in the air gap in the radial direction (the subscript "f" is for field). If permanent magnets like those shown in Figure 11.2

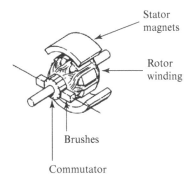

Stator
magnets

Rotor
winding

Brushes

Commutator

FIGURE 11.2 Exploded view of a DC motor. Source: Electro-Craft Corporation.

are used, the air gap flux density established by the stator remains constant. A field winding whose current can be varied can be used to achieve an additional degree of control over the air gap flux density.

As shown in Figure 11.2, the rotor slots contain a winding, called the armature winding, which handles electrical power for conversion to (or from) mechanical power at the rotor shaft. In addition, there is a commutator affixed to the rotor. On its outer surface, the commutator contains copper segments, which are electrically insulated from each other by means of mica or plastic. The coils of the armature winding are connected to these commutator segments so that a stationary DC source can supply voltage and current to the rotating commutator by means of stationary carbon brushes that rest on top of the commutator. The wear due to the mechanical contact between the commutator and the brushes requires periodic maintenance, which is the main drawback of DC machines.

In a DC machine, the magnitude and the direction of the electromagnetic torque depend on the armature current i_a. Therefore, in a permanent-magnet DC machine, the electromagnetic torque produced by the machine is linearly related to a machine torque constant k_T, which is unique to a given machine and is specified in its data sheets,

$$T_{em} = k_T\, i_a. \tag{11.1}$$

Similarly, the induced emf e_a depends only on the rotational speed ω_m, and can be related to it by a voltage constant k_E, which is unique to a given machine and specified in its data sheets,

$$e_a = k_E\, \omega_m. \tag{11.2}$$

Equating the mechanical power ($\omega_m T_{em}$) to the electrical power ($e_a i_a$), the torque constant k_T and the voltage constant k_E are exactly the same numerically in MKS units,

$$k_T = k_E. \tag{11.3}$$

The direction of the armature current i_a determines the direction of currents through the conductors. Therefore, the direction of the electromagnetic torque produced by the machine also depends on the direction of i_a. This explains how a DC machine, while rotating in a forward or reverse direction, can be made to go from motoring mode (where the speed and torque are in the same direction) to its generator mode (where the speed and torque are in the opposite direction) by reversing the direction of i_a.

Applying a reverse-polarity DC voltage to the armature terminals makes the armature current flow in the opposite direction. Therefore, the electromagnetic torque is reversed, after some time reversing the direction of rotation and the polarity of induced emfs in conductors, which depends on the direction of rotation.

The above discussion shows that a DC machine can easily be made to operate as a motor or as a generator in the forward or in the reverse direction of rotation.

It is convenient to discuss a DC machine in terms of its equivalent circuit, presented in Figure 11.3, which shows the conversion between electrical and mechanical power. The armature current i_a produces the electromagnetic torque $T_{em}(= k_T i_a)$ necessary to rotate the mechanical load at the desired speed. Across the armature terminals, the rotation at the speed of ω_m induces a voltage, called the back-emf $e_a(= k_E \omega_m)$, represented by a dependent voltage source.

The applied voltage v_a from the PPU overcomes the back-emf e_a and causes the current i_a to flow. Recognizing that there is a voltage drop across both the armature winding resistance R_a (which includes the voltage drop across the carbon brushes) and the armature winding inductance L_a, we can write the equation of the electrical side as

$$v_a = e_a + R_a\, i_a + L_a \frac{di_a}{dt}. \tag{11.4}$$

On the mechanical side, the electromagnetic torque produced by the motor overcomes the mechanical-load torque T_L to produce acceleration:

$$\frac{d\omega_m}{dt} = \frac{1}{J_{eq}}(T_{em} - T_L), \tag{11.5}$$

where $J_{eq}(= J_m + J_L)$ is the total effective value of the combined inertia of the DC machine and the mechanical load.

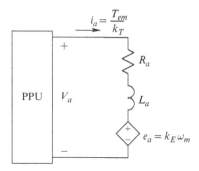

FIGURE 11.3 DC motor equivalent circuit.

Note that the electric and the mechanical systems are coupled. The torque T_{em} in the mechanical system (Equation 11.1) depends on the electrical current i_a. The back-emf e_a in the electrical system (Equation 11.2) depends on the mechanical speed ω_m. The electrical power absorbed from the electrical source by the motor is converted into mechanical power and vice versa. In a DC steady state, with a voltage V_a applied to the armature terminals and a load torque $T_L(= T_{em})$ being supplied to the load, the inductance L_a in the equivalent circuit of Figure 11.3 does not play a role. Hence, in Figure 11.3,

$$I_a = \frac{T_{em}(= T_L)}{k_T} \tag{11.6}$$

$$\omega_m = \frac{E_a}{k_E} = \frac{V_a - R_a I_a}{k_E} = \frac{V_a - R_a (T_{em}/k_T)}{k_E}. \tag{11.7}$$

11.2.1.1 Requirements Imposed by DC Machines on the PPU

Based on Equations 11.6 and 11.7, the steady-state torque-speed characteristics for various values of V_a are plotted in Figure 11.4a. Neglecting the voltage drop across the armature resistance in Equation 11.7, the terminal voltage V_a varies linearly with the speed ω_m, as plotted in Figure 11.4b. At low values of speed, the voltage drop across the armature resistance can be a significant component of the terminal voltage, and hence the relationship in Figure 11.4a in this region is shown dotted.

In summary, in DC-motor drives, the voltage rating of the power-processing unit is dictated by the maximum speed, and the current rating is dictated by the maximum load torque being supplied.

11.2.2 Permanent-Magnet AC Machines

Sinusoidal-waveform, permanent-magnet AC (PMAC) drives are used in applications where high efficiency and high power density are required in controlling the speed and position. In trade literature, they are also called "brushless DC" drives. We will examine these machines for speed and position control applications, usually in small (<10 kW) power ratings, where these drives have three-phase AC stator windings, and the rotor

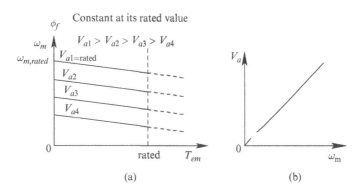

FIGURE 11.4 (a) Torque-speed characteristics, and (b) V_a versus ω_m.

has DC excitation in the form of permanent magnets. In such drives, the stator windings of the machine are supplied by controlled currents, which require a closed-loop operation, as described later in this section.

We will consider 2-pole machines, like the one shown schematically in Figure 11.5a. This analysis can be generalized to p-pole machines where $p > 2$. The stator contains three-phase, wye-connected windings, each of which produces a sinusoidally distributed flux-density distribution in the air gap when supplied by a current.

The permanent-magnet pole pieces mounted on the rotor surface are shaped to ideally produce a sinusoidally distributed flux density in the air gap. The rotor flux-density distribution (represented by a vector $\vec{B}_r V_a$) peaks at an axis that is at an angle $\theta_m(t)$ with respect to the a-axis (the phase axis of winding a), as shown in Figure 11.5b. As the rotor turns, the entire rotor-produced flux density distribution in the air gap rotates with it and "cuts" the stator-winding conductors and produces emf in phase windings that are sinusoidal functions of time. In the AC steady state, these voltages can be represented by phasors.

Considering phase-a as the reference in Figure 11.5b, the induced voltage in it due to the rotor flux cutting it can be expressed as

$$\bar{E}_{ma} = E_{rms} \angle 0°. \tag{11.8}$$

It should be noted that by Faraday's law ($e = d\lambda / dt$, where λ is the flux-linkage of the coil), the induced voltage in phase-a peaks when the rotor-flux peak in Figure 11.5a is pointing downward, 90° before it reaches the phase-a magnetic axis. Since the permanent magnets on the rotor produce a constant amplitude flux, the RMS magnitude of the induced voltage in each stator phase is linearly related to the rotational speed ω_m by a per-phase voltage-constant $k_{E,phase}$,

$$E_{rms} = k_{E,phase} \, \omega_m. \tag{11.9}$$

An important characteristic of the machines under consideration is that they are supplied through a current-regulated power-processing unit, which controls the currents

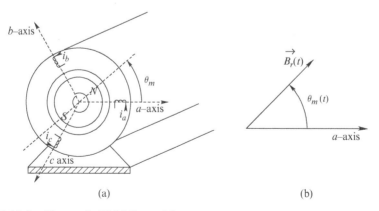

(a)   (b)

FIGURE 11.5   Two-pole PMAC machine.

i_a, i_b, and i_c supplied to the stator at any instant of time. To optimize such that the maximum torque-per-ampere is produced, each phase current in AC steady-state is controlled in phase with the induced voltage. Therefore, with the voltage expressed in Equation 11.8, the current in AC steady state is

$$\overline{I}_a = I_{\text{rms}}\angle 0° \tag{11.10}$$

In AC steady state, accounting for all three phases, the input electric power, supplied by the current in opposition to the induced back-emf, equals the mechanical output power. Using Equations 11.8 through 11.10,

$$T_{em}\omega_m = 3(\underbrace{k_{E,phase}\omega_m}_{E_{rms}})I_{rms}. \tag{11.11}$$

The torque contribution of each phase is $T_{em}/3$, so from Equation 11.11,

$$T_{em,phase} = \frac{T_{em}}{3} = k_{E,phase}\,I_{\text{rms}}, \tag{11.12}$$

where the constant that relates the RMS current to the per-phase torque is the per-phase torque constant $k_{T,phase}$, and similar to that in DC machines, in MKS units,

$$k_{T,phase} = k_{E,phase}. \tag{11.13}$$

At this point, we should note that PMAC drives constitute a class, which we will call *self-synchronous* motor drives, where the term *"self"* is added to distinguish these machines from the conventional synchronous machines. The reason for this is as follows: in PMAC drives, the stator phase currents are synchronized to the mechanical position of the rotor such that, for example, the current into phase-a, in order to be in phase with the induced emf, peaks when the rotor-flux peak is pointing downward, 90° before it reaches the phase-a magnetic axis. This explains the necessity for the rotor-position sensor, as shown in the block diagram of such drives in Figure 11.6.

As shown in the block diagram of Figure 11.6, the task of the controller is to enable the power-processing unit to supply the desired currents to the PMAC motors.

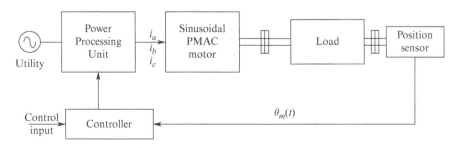

FIGURE 11.6 Block diagram of a PMAC machine.

The reference torque signal is generated from the outer speed and position loops. The rotor position θ_m is measured by the position sensor connected to the shaft. Knowing the torque constant $k_{T,phase}$ allows us to calculate the RMS value of the reference current from Equation 11.12. Knowing the current RMS value and θ_m allows the reference currents i_a, i_b, and i_c for the three phases to be calculated, which the PPU delivers at any instant of time.

The electromagnetic torque acts on the mechanical system connected to the rotor, and the resulting speed ω_m can be obtained from the equation below:

$$\frac{d\omega_m}{dt} = \frac{T_{em} - T_L}{J_{eq}} \Rightarrow \omega_m(t) = \omega_m(0) + \frac{1}{J_{eq}} \int_0^t (T_{em} - T_L) \cdot d\tau, \qquad (11.14)$$

where J_{eq} is the combined motor-load inertia and T_L is the load torque, which may include friction. The rotor position $\theta_m(t)$ is

$$\theta_m(t) = \theta_m(0) + \int_0^t \omega_m(\tau) \cdot d\tau, \qquad (11.15)$$

where $\theta_m(0)$ is the rotor position at time $t = 0$. Similar to a DC machine, it is convenient to discuss a PMAC machine in terms of its equivalent circuit of Figure 11.7, which shows the conversion between electrical and mechanical power.

In the AC steady state, using phasors, the current $\bar{I}_a$ is ensured by the feedback control to be in phase with the phase-a induced voltage $\bar{E}_{ma}$. The phase currents produce the total electromagnetic torque necessary to rotate the mechanical load at a speed ω_m. The induced back-emf $\bar{E}_{ma} = E_{rms}\angle 0°$, whose RMS magnitude is linearly proportional to the speed of rotation ω_m, is represented by a dependent voltage source.

The applied voltage $\bar{V}_a$ in Figure 11.7a overcomes the back-emf $\bar{E}_{ma}$ and causes the current $\bar{I}_a$ to flow. The frequency of the phasors in hertz equals $\omega_m/(2\pi)$ in a 2-pole PMAC machine. There is a voltage drop across both the per-phase stator winding resistance R_s (neglected here) and a per-phase inductance L_s, which is the sum of the leakage inductance $L_{\ell s}$ caused by the leakage flux of the stator winding and L_m due to the effect of the combined flux produced by the currents flowing in the stator phases. The phasor diagram, neglecting R_s, is shown in Figure 11.7b.

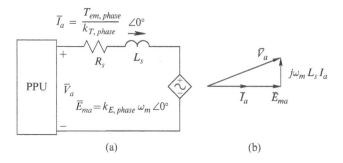

(a)   (b)

FIGURE 11.7   Equivalent circuit diagram and the phasor diagram of PMAC (2-pole).

11.2.2.1 Requirements Imposed by PMAC Machines on the PPU

In PMAC machines, the speed is independent of the electromagnetic torque developed by the machine, as shown in Figure 11.8a, and depends on the frequency of voltages and currents applied to the stator phases of the machine.

In the per-phase equivalent circuit of Figure 11.7a and the phasor diagram of Figure 11.7b, the back-emf is proportional to the speed ω_m. Similarly, since the electrical frequency is linearly related to ω_m, the voltage drop across L_s is also proportional to ω_m. Therefore, neglecting the per-phase stator winding resistance R_s, in the phasor diagram of Figure 11.7b, the voltage phasors are all proportional to ω_m, requiring that the per-phase voltage magnitude that the power-processing unit needs to supply is proportional to the speed ω_m, as plotted in Figure 11.8b.

At very low speeds, this voltage-speed relationship, shown dotted in Figure 11.8b, is approximate, where a higher voltage, called the voltage boost, is needed to overcome the voltage drop across the stator winding resistance R_s; this voltage drop becomes significant at higher torque loading and hence cannot be neglected.

In summary, in PMAC-motor drives, similar to DC drives, the voltage rating of the power-processing unit is dictated by the maximum speed, and the current rating is dictated by the maximum load torque being supplied.

11.2.3 Induction Machines

Induction motors with squirrel-cage rotors are the workhorses of industry because of their low cost and rugged construction. When operated directly from line voltages (a 50- or 60-Hz utility input at essentially a constant voltage), induction motors operate at a nearly constant speed. However, by means of power electronic converters, it is possible to vary their speed efficiently.

Just as in PMAC motors, the stator of an induction motor consists of three-phase windings distributed in the stator slots. These three windings are displaced by 120° in space with respect to each other, as shown by their axes in Figure 11.9a.

The rotor, consisting of a stack of insulated laminations, has electrically conducting bars of copper or aluminum inserted (molded) through it, close to the periphery in the axial direction. These bars are electrically shorted at each end of the rotor by electrically conducting end-rings, thus producing a conducting cage-like structure, as shown in Figure 11.9b. Such a rotor, called a squirrel-cage rotor, has a low cost and rugged nature.

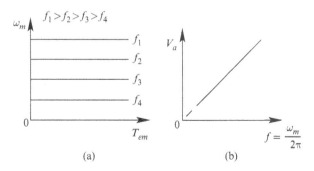

FIGURE 11.8 Torque-speed characteristics and voltage versus frequency in PMAC.

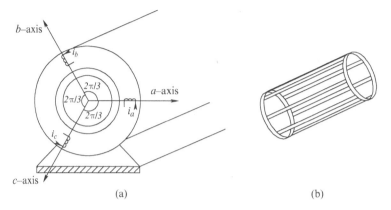

FIGURE 11.9   (a) Three-phase stator; (b) squirrel-cage rotor.

Figure 11.10a shows the stator windings whose voltages are shown in the phasor diagram of Figure 11.10b, where the frequency of the applied line voltages to the motor is f in hertz, and V_{rms} is the magnitude in RMS.

$$\bar{V}_a = V_{rms}\angle 0°, \quad \bar{V}_b = V_{rms}\angle -120°, \quad and \quad \bar{V}_c = V_{rms}\angle -240°. \tag{11.16}$$

By Faraday's law, the applied stator voltages given in Equation 11.16 establish a rotating flux-density distribution in the air gap by drawing magnetizing currents of RMS value I_m, which are shown in Figure 11.10b:

$$\bar{I}_{ma} = I_m\angle -90°, \quad \bar{I}_{mb} = I_m\angle -210°, \quad and \quad \bar{I}_{mc} = I_m\angle -330°. \tag{11.17}$$

As these currents vary sinusoidally with time, the combined flux-density distribution in the air gap, produced by these currents, rotates with a constant amplitude at a synchronous speed ω_{syn}, where

$$\omega_{syn} = 2\pi f \left(\omega_{syn} = \frac{2\pi f}{p/2} \text{for a } p\text{-pole machine}\right) \tag{11.18}$$

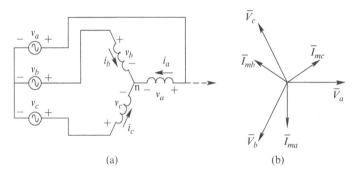

FIGURE 11.10   Induction machine: applied voltages and magnetizing currents.

The rotor in an induction motor turns (due to the electromagnetic torque developed, as will be discussed shortly) at a speed ω_m in the same direction as the rotation of the air gap flux-density distribution, such that $\omega_m < \omega_{syn}$. Therefore, there is a relative speed between the flux-density distribution rotating at ω_{syn} and the rotor conductors at ω_m. This relative speed, that is, the speed at which the rotor is "slipping" with respect to the rotating flux-density distribution, is called the slip speed:

$$\text{slip speed } \omega_{slip} = \omega_{syn} - \omega_m \tag{11.19}$$

By Faraday's law ($e = Blu$), voltages are induced in the rotor bars at the slip frequency due to the relative motion between the flux-density distribution and the rotor, where the slip frequency f_{slip} in terms of the frequency f of the stator voltages and currents is

$$\text{slip frequency } f_{slip} = \frac{\omega_{slip}}{\omega_{syn}} f. \tag{11.20}$$

Since the rotor bars are shorted at both ends, these induced bar voltages cause slip-frequency currents to flow in the rotor bars. The rotor-bar currents interact with the flux-density distribution established by the stator-applied voltages, and the result is a net electromagnetic torque T_{em} in the same direction as the rotor's rotation.

The sequence of events in an induction machine to meet the load torque demand is as follows: At essentially no load, an induction machine operates nearly at the synchronous speed that depends on the frequency of applied stator voltages (Equation 11.18). As the load torque increases, the motor slows down, resulting in a higher value of slip speed. Higher slip speed results in higher voltages induced in the rotor bars and hence higher rotor-bar currents. Higher rotor-bar currents result in a higher electromagnetic torque to satisfy the increased load-torque demand.

Neglecting second-order effects, the air gap flux is totally determined by the applied stator voltages. Hence, the air gap flux produced by the rotor-bar currents is nullified by the additional currents drawn by the stator windings, which are in addition to the magnetizing currents in Equation 11.17.

In this balanced three-phase sinusoidal steady-state analysis, we will neglect second-order effects such as the stator winding resistance and leakage inductance and the rotor circuit leakage inductance. As shown in Figure 11.11a for phase-a in this

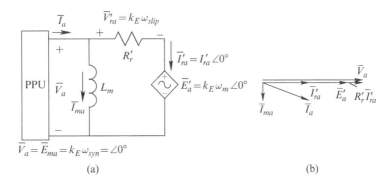

(a) (b)

FIGURE 11.11 Induction motor equivalent circuit and phasor diagram.

per-phase circuit, $\bar{V}_a$ and, similarly for the other two phases, are applied at a frequency f, which results in magnetizing currents to establish the air gap flux-density distribution in the air gap.

These magnetizing currents are represented as flowing through a magnetizing inductance L_m. The voltage magnitude and the frequency are such that the magnitude of the flux-density distribution in the air gap is at its rated value, and this distribution rotates counterclockwise at the desired ω_{syn} (Equation 11.18) to induce a back-emf in the stationary stator phase windings such that

$$\bar{V}_a = \bar{E}_{ma} = k_{E,phase}\omega_{syn}\angle 0°, \tag{11.21}$$

where $k_{E,phase}$ is the machine per-phase voltage-constant.

Next, we will consider the effect of the rotor-bar currents. The rotor-bar currents result in additional stator currents (in addition to the magnetizing current), which can be represented on a per-phase basis by a current $\bar{I}'_{ra}$ in-phase with the applied voltage (since the rotor-circuit leakage inductance is ignored), as shown in Figure 11.11a,

$$\bar{I}'_{ra} = I'_{ra}\angle 0°. \tag{11.22}$$

Note that the torque on the stator is equal in magnitude and opposite in direction to the torque on the rotor. The flux-density distribution (although produced differently than in PMAC machines) interacts with these stator currents, just like in PMAC machines, and hence the per-phase torque constant equals the per-phase voltage constant,

$$k_{T,phase} = k_{E,phase}. \tag{11.23}$$

The per-phase torque can be expressed as

$$T_{em,phase} = k_{T,phase}\, I'_{ra} \tag{11.24}$$

This torque at a mechanical speed ω_m results in per-phase electromagnetic power that gets converted to mechanical power, where using Equations 11.23 and 11.24,

$$P_{em,phase} = \omega_m T_{em,phase} = \omega_m\, k_{E,phase}\, I'_{ra}. \tag{11.25}$$

In Equation 11.25, $(\omega_m\, k_{E,phase})$ can be considered the back-emf $\bar{E}'_a$, just like in the DC and the PMAC machines, as shown in Figure 11.11a,

$$\bar{E}'_a = \omega_m\, k_{E,phase}\angle 0° \tag{11.26}$$

In the rotor circuit, the voltages induced depend on the slip speed ω_{slip} and overcome the IR voltage drop in the rotor bar resistances. The rotor bar resistances, the voltage drop, and the power losses in them are represented in the per-phase equivalent circuit of Figure 11.11a by a voltage drop across an equivalent resistance R'_r. Using Kirchhoff's voltage law in Figure 11.11a,

$$\bar{V}'_{ra} = \underbrace{k_{E,phase}\omega_{syn}\angle 0°}_{\bar{E}_{ma}} - \underbrace{k_{E,phase}\omega_m\angle 0°}_{\bar{E}'_a} = k_{E,phase}\,\omega_{slip}\angle 0°. \tag{11.27}$$

Hence,

$$I'_{ra} = \left(\frac{k_{E,phase}}{R'_r}\right)\omega_{slip}. \tag{11.28}$$

Using Equations 11.23, 11.25, and 11.28, the combined torque of all three phases is

$$T_{em} = \frac{P_{em}}{\omega_m} = 3\underbrace{\frac{k^2_{T,phase}}{R'_r}}_{k_{T,\omega slip}}\omega_{slip} \tag{11.29}$$

where $k_{T,\omega_{slip}}$ in the above equation is a machine constant that shows that the torque produced is linearly proportional to the slip speed. Using Equations 11.19 and 11.29,

$$\omega_m = \omega_{syn} - \frac{T_{em}}{k_{T,\omega_{slip}}}. \tag{11.30}$$

11.2.3.1 Requirements Imposed by Induction Machines on the PPU

Based on Equation 11.30, the torque-speed characteristics are shown in Figure 11.12 for various applied frequencies to the stator. Based on Equation 11.21, the stator voltage as a function of frequency is shown in Figure 11.13 to maintain the peak of the flux-density distribution at its rated value. The relationship between the voltage and the synchronous speed or f shown in Figure 11.13, is approximate. A substantially higher voltage than indicated (shown dotted) is needed at very low frequencies where the voltage drop across the stator winding resistance R_s becomes substantial at higher torque loading and hence cannot be neglected.

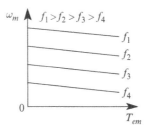

FIGURE 11.12 Induction motor torque-speed characteristics.

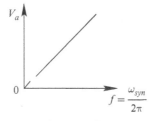

FIGURE 11.13 Induction motor voltage vs frequency.

In summary, in induction motor drives, similar to DC and PMAC drives, the voltage rating of the power-processing unit is dictated by the maximum speed, and the current rating is dictated by the maximum load torque being supplied.

11.3   UNINTERRUPTIBLE POWER SUPPLIES (UPS)

Uninterruptible power supplies (UPS) are used to provide power to critical loads in industry, business, and medical facilities, to which power should be available continuously, even during momentary utility power outages. The Computer Business Equipment Manufacturers Association (CBEMA) has specifications that show the voltage tolerance envelope as a function of time. The ITI-CBEMA curve shows that the UPS for critical loads is needed, not just for complete power outages but also for "swells" and "sags" in the equipment voltage due to disturbances on the utility grid. In considering the synthesis of a low-frequency AC from a DC voltage, UPS can be considered as a special case of AC motor drives.

In a UPS, storage means, generally a battery bank, is used to shield critical loads from voltage disturbances, as shown in the block diagram of Figure 11.14.

Therefore, there is an intermediate DC link. Normally, the power to the load is provided through the two converters, where the utility-side converter also keeps the batteries charged. In the event of power-line outages, energy stored in the batteries allows load power to be supplied continuously. The function of the load-side inverter is to produce, from the DC source, AC voltages similar to that in AC motor drives, except it is a special case where the output voltage has a constant specified magnitude and frequency (e.g., 115 V RMS, and 60 Hz). The load may be single-phase or three-phase. The voltage and current ratings of the UPS converters are dictated by the load that is being supplied.

11.4   UTILITY APPLICATIONS OF SWITCH-MODE POWER ELECTRONICS

Thyristor-based applications of power electronics in utility systems are well established in the form of HVDC transmission, excitation control of generators, switched capacitors for static var control, and so on. In addition to these applications based on thyristor converters discussed in Chapter 14, there are growing applications of switch-mode power electronics. These include voltage-link HVDC systems, discussed in Chapter 14, and static compensators (STATCOM) for var control. Other applications include series-voltage injection for controlling the flow of power and for compensating voltage sags and swells beyond the limits specified by the CBEMA curve of Figure 11.13.

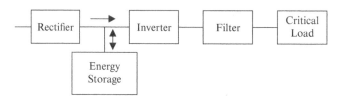

FIGURE 11.14   Block diagram of UPS.

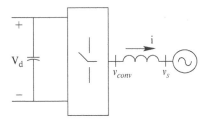

FIGURE 11.15 Interaction of the switch-mode converter with the AC utility system.

All the above applications of switch-mode power electronics are based on the fact that in a switched-mode converter, shown in Figure 11.15 as a block diagram with the DC voltage at one side forming the voltage port, a voltage of appropriate amplitude, frequency, and phase can be synthesized at the current port. This converter can have single- or three-phase outputs, and the power flow can be bidirectional. The utility system is shown at the right with a voltage v_s.

The per-phase AC side is shown in Figure 11.16a by means of the fundamental-frequency quantities, where $\bar{V}_{conv}$ is the voltage synthesized by the converter, and $\bar{V}_S$ is the utility voltage. These two AC voltage sources of the same frequency are connected through a reactance X in series. The phasor diagram for the variables in Figure 11.16a is shown in Figure 11.16b.

The following discussion shows that the real power P flows "downhill" on the phase angles of the voltages, and not their magnitudes, whereas the reactive power Q flows based on the magnitudes of the two voltages.

In the circuit of Figure 11.16a,

$$\bar{I} = \frac{\bar{V}_{conv} - \bar{V}_s}{jX}. \tag{11.31}$$

At the utility end, the complex power can be written as

$$S_s = P_s + jQ_s = V_s\bar{I}*. \tag{11.32}$$

Using the complex conjugate from Equation 11.31 into Equation 11.32, and selecting the utility voltage as the reference phasor $\bar{V}_s = V_s\angle0$ where $\bar{V}_{conv} = V_{conv}\angle\delta$,

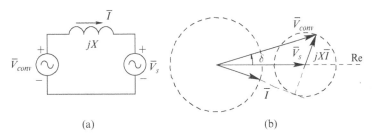

(a)   (b)

FIGURE 11.16 Per-phase equivalent circuit and the phasor diagram.

$$P_s + jQ_s = V_s \left(\frac{V_{conv}\angle(-\delta) - V_s}{-jX} \right) = \frac{V_{conv}V_s \sin\delta}{X} + j\left(\frac{V_{conv}V_s \cos\delta - V_s^2}{X} \right) \quad (11.33)$$

$$\therefore \qquad P_s = \frac{V_{conv}V_s}{X}\sin\delta, \qquad (11.34)$$

which is the same as the sending end power P_{conv} assuming no transmission-line losses. And

$$Q_s = \frac{V_{conv}V_s \cos\delta}{X} - \frac{V_s^2}{X}. \qquad (11.35)$$

If the power transfer from the converter to the utility system is zero, then, from Equation 11.34, $\sin\delta$ and, hence the angle δ is equal to zero. Under this condition, from Equation 11.35,

$$Q_s = \frac{V_{conv}V_s}{X} - \frac{V_s^2}{X} = \frac{V_s}{X}\left(V_{conv} - V_s \right) \quad (\text{if } P_s = 0), \qquad (11.36)$$

where the reactive power is supplied by the converter, as if it were a capacitor, to the utility system.

Equations 11.34 and 11.35 together show that by controlling $\bar{V}_{conv}$, in amplitude and phase, it is possible to control both the real and the reactive powers, as shown by the circular trajectories of voltage and current phasors in Figure 11.16b. This is what's done in HVDC systems. Equation 11.36 shows that when the real power transfer is zero ($\delta = 0$), by controlling the amplitude V_{conv} in comparison to the utility-grid voltage magnitude, it is possible to control Q in magnitude and polarity, that is, supply vars to the utility grid as a shunt-connected capacitor would or absorb it from the utility grid as a shunt-connected inductor would. This is what's done in static var compensators (STATCOM). These applications, such as HVDC and STATCOM, are further discussed in Chapter 14.

REFERENCE

1. N. Mohan, *Electric Drives: An Integrative Approach* (New York: John Wiley & Sons, 2011).

PROBLEMS

Mechanical Systems

11.1 A constant torque of 5 Nm is applied to an unloaded motor at rest at time $t = 0$. The motor reaches a speed of 1800 rpm in 5 s. Assuming the damping to be negligible, calculate the motor inertia.

11.2 In an electric motor drive, the combined motor-load inertia is $J_{eq} = 5\times10^{-3}\ \text{kg}\cdot m^2$. The load torque is $T_L = 0.05\text{Nm}$. Plot the electromagnetic torque required from the motor to bring the system linearly from rest to a speed of 100 rad/s in 5 s and then maintain that speed.

11.3 Calculate the power required in Problem 11.2.

DC Motors

11.4 A permanent-magnet DC motor has the following parameters: $R_a = 0.3\,\Omega$ and $k_E = k_T = 0.5$ in MKS units. For a torque of up to 10 Nm, plot its steady-state torque-speed characteristics for the following values of V_a: 100 V, 75 V, and 50 V.

11.5 Consider the DC motor of Problem 11.4 whose moment of inertia $J_m = 0.02\,\text{kg} \cdot m^2\, R_a = 0.3\,\Omega$. Its armature inductance L_a can be neglected for slow changes. The motor is driving a load of inertia $J_L = 0.04\,\text{kg} \cdot m^2$. The steady-state operating speed is 400 rad/s. Calculate and plot the terminal voltage $v_a(t)$ that is required to bring this motor to a halt as quickly as possible, without exceeding the armature current of 12 A.

Permanent-Magnet AC Motors

11.6 In a three-phase, 2-pole PMAC motor, the torque constant $k_{T,phase} = 0.5\,\text{Nm}/A$. Calculate the phase currents if the motor is to produce a counterclockwise torque of 5 Nm.

11.7 In a 2-pole, three-phase PMAC motor drive, the torque constant $k_{T,phase}$ and the voltage constant $k_{E,1-phase}$ are 0.5 in MKS units. The synchronous inductance is 20 mH (neglect the winding resistance). This motor is supplying a torque of 3 Nm at a speed of 3,000 rpm in a balanced sinusoidal steady state. Calculate the per-phase voltage across the power-processing unit as it supplies controlled currents to this motor.

Induction Motors

11.8 Consider an induction machine that has 2 poles and is supplied by a rated voltage of 208 V (line-to-line, RMS) at the frequency of 60 Hz. It is operating in steady state and is loaded to its rated torque. Neglect the stator leakage impedance and the rotor leakage flux. The per-phase magnetizing current is 4.0 A (RMS). The current drawn per-phase is 10 A (RMS) and is at an angle of 23.56 degrees (lagging). Calculate the per-phase current if the mechanical load decreases so that the slip speed is one-half that of the rated case.

11.9 In Problem 11.8, the rated speed (while the motor supplies its rated torque) is 3480 rpm. Calculate the slip speed ω_{slip} and the slip frequency f_{slip} of the currents and voltages in the rotor circuit.

11.10 In Problem 11.9, the rated torque supplied by the motor is 8 Nm. Calculate the torque constant, which linearly relates the torque developed by the motor to the slip speed.

11.11 A three-phase, 60-Hz, 4-pole, 440-V (line-line, RMS) induction motor drive has a full-load (rated) speed of 1750 rpm. The rated torque is 40 Nm. Keeping the air gap flux-density peak constant at its rated value, (a) plot the torque-speed characteristics (the linear portion) for the following values of the frequency f: 60 Hz, 45 Hz, 30 Hz, and 15 Hz. (b) This motor is supplying a load whose torque demand increases linearly with speed, such that it equals the rated torque of the motor at the rated motor speed. Calculate the speeds of operation at the four values of frequency in part (a).

11.12 In the motor drive of Problem 11.11, the induction motor is such that while applying the rated voltages and loaded to the rated torque, it draws 10.39 A (RMS) per-phase at a power factor of 0.866 (lagging). $R_s = 1.75\Omega$. Calculate the voltages corresponding to the four values of the frequency f to maintain $\widehat{B}_{ms} = \widehat{B}_{ms,rated}$.

12

SYNTHESIS OF DC AND LOW-FREQUENCY SINUSOIDAL AC VOLTAGES FOR MOTOR DRIVES, UPS, AND POWER SYSTEMS APPLICATIONS

12.1 INTRODUCTION

The importance of power electronics applications for motor drives (AC and DC), UPS, and in power systems was described in Chapter 11. In many of these applications, the voltage-link structure of Figure 12.1 is used, where our emphasis will be to discuss how the load-side converter, with the DC voltage as input, synthesizes DC or low-frequency sinusoidal voltage outputs. Functionally, this converter operates as a linear amplifier, amplifying a control signal, DC in case of DC motor drives, and AC

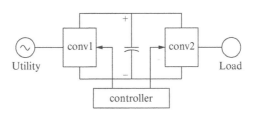

FIGURE 12.1 Voltage-link system.

Power Electronics A First Course: Simulations and Laboratory Implementations, Second Edition.
Ned Mohan and Siddharth Raju.
© 2023 John Wiley & Sons, Inc. Published 2023 by John Wiley & Sons, Inc.
Companion Website: www.wiley.com/go/mohan/powerelectronics2e

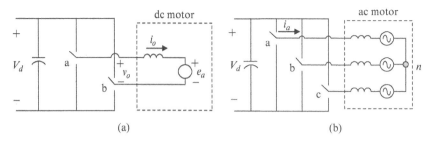

FIGURE 12.2   Converters for DC and AC motor drives.

in case of AC motor drives, UPS, and other utility-related applications. The power flow through this converter should be reversible.

These converters consist of bidirectional switching power-poles, discussed in Chapter 3, two in the case of DC motor drives and single-phase AC applications, and three in the case of AC motor drives and three-phase applications. These are shown for DC and AC motor drives in Figures 12.2a and 12.2b, respectively.

12.2   BIDIRECTIONAL SWITCHING POWER-POLE AS THE BUILDING BLOCK

In buck and boost DC-DC converters, discussed in Chapter 3, implementation of the switching power-pole by one transistor and one diode dictates the instantaneous current flow to be unidirectional. However, as shown in Figure 12.3a, combining the switching power-pole implementations of buck and boost converters, where the two transistors are switched by complementary signals, allows a continuous bidirectional power and current capability. In such a bidirectional switching power-pole, the positive inductor current i_L as shown in Figure 12.3b, represents a buck-mode of operation, where only the transistor and the diode associated with the buck converter take part. The transistor conducts during $q = 1$, and the diode conducts during $q = 0$. Similarly, as shown in Figure 12.3c, the negative inductor current represents a boost-mode of operation, where only the transistor and the diode associated with the boost converter take part. The transistor conducts during $q = 0$ ($q^- = 1$), and the diode conducts during $q = 1$ ($q^- = 0$).

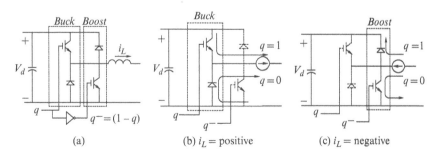

FIGURE 12.3   Bidirectional power flow through a switching power-pole.

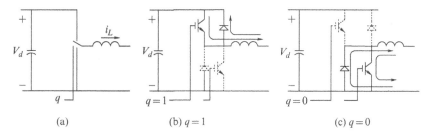

FIGURE 12.4 Bidirectional switching power-pole.

Figures 12.3b and 12.3c show that the combination of devices in Figure 12.3a renders it to be a switching power-pole that can carry i_L in either direction. This is shown as an equivalent switch in Figure 12.4a that is effectively in the "up" position when $q = 1$, as shown in Figure 12.4b, and in the "down" position when $q = 0$, as shown in Figure 12.4c, regardless of the direction of i_L.

The bidirectional switching power-pole of Figure 12.4a is repeated in Figure 12.5a for pole-a, with its switching signal identified as q_a. In response to the switching signal, it behaves similarly to the switching power-pole in Chapter 3: "up" when $q_a = 1$ and otherwise "down." Therefore, its switching-cycle-averaged representation is also an ideal transformer, shown in Figure 12.5b, with a turns ratio $1:d_a(t)$.

The switching-cycle-averaged values of the variables at the voltage port and the current port in Figure 12.5b are related by $d_a(t)$ as follows:

$$\bar{v}_{aN} = d_a V_d \tag{12.1}$$

$$\bar{i}_{da} = d_a \bar{i}_a. \tag{12.2}$$

We should note that, ideally, unlike switching power-poles with a single transistor, discussed in Chapter 3, no discontinuous-conduction mode exists in a bidirectional switching pole of Figure 12.5a.

12.2.1 Pulse-Width Modulation (PWM) of the Bidirectional Switching Power-Pole

The voltage of a switching power-pole at the current port is always of positive polarity. However, the output voltages of converters in Figure 12.2 for motor drives and other AC applications must be reversible in polarity. This is achieved by introducing a common-mode voltage in each power pole as discussed below and taking the differential output between the power poles.

To obtain the desired switching-cycle-averaged voltage $\bar{v}_{aN}$ in Figure 12.5, which includes the common-mode voltage, requires the following power-pole duty ratio from Equation 12.1:

$$d_a = \frac{\bar{v}_{aN}}{V_d}, \tag{12.3}$$

where V_d is the DC-bus voltage. To obtain the switching signal q_a to deliver this duty ratio, a control voltage $v_{cntrl,a}$ is compared with a triangular-shaped carrier waveform

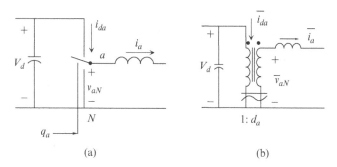

(a)   (b)

FIGURE 12.5 Switching-cycle-averaged representation of the bidirectional power pole.

of the switching frequency f_s and amplitude $\hat{V}_{tri}$, as shown in Figure 12.6. Because of symmetry, only $T_s / 2$, one-half of the switching time period, needs to be considered. The switching-signal $q_a = 1$ if $v_{cntrl,a} > v_{tri}$ and is otherwise 0. Therefore in Figure 12.6,

$$v_{cntrl,a} = d_a \hat{V}_{tri}. \qquad (12.4)$$

The switching-cycle-averaged representation of the switching power-pole in Figure 12.7a is shown by a controllable turn-ratio ideal transformer in Figure 12.7b, where the switching-cycle-averaged representation of the duty-ratio control is in accordance with Equation 12.4.

The reason for selecting a triangular carrier-signal waveform, as opposed to a ramp signal in DC-DC converters of Chapter 3, is that it minimizes the harmonic content in the output voltage waveform for a given frequency with which the converter devices are switched. The Fourier spectrum of the switching waveform v_{aN} is shown in Figure 12.8, which depends on the nature of the control signal. If the control voltage is DC, the output voltage has harmonics at the multiples of the switching frequency,

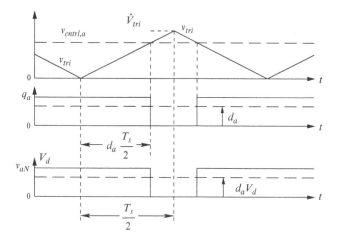

FIGURE 12.6 Waveforms for PWM in a switching power-pole.

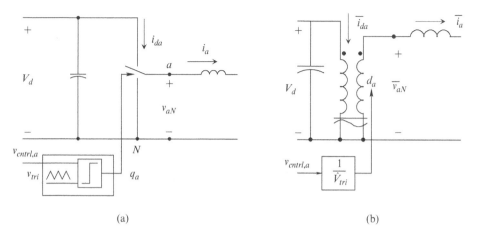

(a)   (b)

FIGURE 12.7   Switching power-pole and its duty-ratio control.

that is at f_s, $2f_s$, and so on, as shown in Figure 12.8a. If the control voltage varies at a low frequency f_1, as in AC motor drives and UPS, then the harmonics of significant magnitudes appear in the side bands of the switching frequency and its multiples, as shown in Figure 12.8b, where

$$f_h = k_1 f_s + \underbrace{k_2 f_1}_{sidebands} \, ,$$ (12.5)

in which k_1 and k_2 are constants that can take on values 1, 2, 3, and so on. Some of these harmonics associated with each power pole are canceled from the converter output voltages, where two or three of such power poles are used.

In the power pole shown in Figure 12.7, the output voltage v_{aN} and its switching-cycle-averaged $\bar{v}_{aN}$ are limited between 0 and V_d. To obtain an output voltage $\bar{v}_{an}$ (where "n" may be a fictitious node) that can become both positive and negative, a

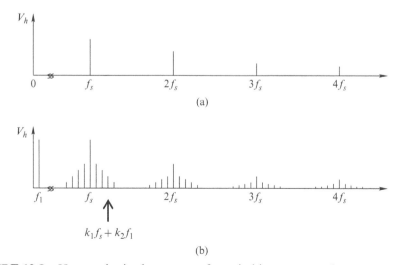

FIGURE 12.8   Harmonics in the output of a switching power-pole.

common-mode offset $\bar{v}_{com}$ is introduced in each power pole so that the pole output voltage is

$$\bar{v}_{aN} = \bar{v}_{com} + \bar{v}_{an}, \qquad (12.6)$$

where $\bar{v}_{com}$ allows $\bar{v}_{an}$ to become both positive and negative around the common-mode voltage $\bar{v}_{com}$. In the differential output, when two or three power poles are used, the common-mode voltage gets eliminated.

12.3 CONVERTERS FOR DC MOTOR DRIVES $(-V_d < \bar{v}_o < V_d)$

Converters for DC motor drives consist of two power poles, as shown in Figure 12.9a, where

$$\bar{v}_o = \bar{v}_{aN} - \bar{v}_{bN}, \qquad (12.7)$$

and $\bar{v}_o$ can assume both positive and negative values. Since the output voltage is desired to be in a full range, from $-V_d$ to $+V_d$, pole-a is assigned to produce $\bar{v}_o / 2$, and pole-b is assigned to produce $-\bar{v}_o / 2$ toward the output:

$$\bar{v}_{an} = \frac{\bar{v}_o}{2} \quad \text{and} \quad \bar{v}_{bn} = -\frac{\bar{v}_o}{2}, \qquad (12.8)$$

where "n" is a fictitious node, as shown in Figure 12.9a, chosen to define the contribution of each pole toward $\bar{v}_o$.

To achieve equal excursions in positive and negative values of the switching-cycle-averaged output voltage, the switching-cycle-averaged common-mode voltage in each pole is chosen to be one-half the DC-bus voltage,

$$\bar{v}_{com} = \frac{V_d}{2}. \qquad (12.9)$$

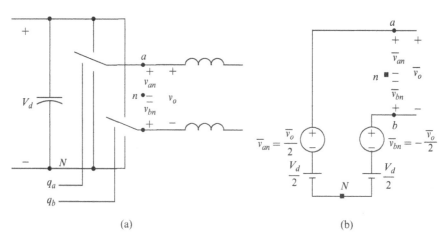

(a)   (b)

FIGURE 12.9 Converter for DC motor drive.

Therefore, from Equation 12.6,

$$\bar{v}_{aN} = \frac{V_d}{2} + \frac{\bar{v}_o}{2} \quad \text{and} \quad \bar{v}_{bN} = \frac{V_d}{2} - \frac{\bar{v}_o}{2}. \tag{12.10}$$

The switching-cycle-averaged output voltages of the power poles and the converter are shown in Figure 12.9b. From Equations 12.3 and 12.10,

$$d_a = \frac{1}{2} + \frac{1}{2}\frac{\bar{v}_o}{V_d} \quad \text{and} \quad d_b = \frac{1}{2} - \frac{1}{2}\frac{\bar{v}_o}{V_d}, \tag{12.11}$$

and from Equation 12.11,

$$\bar{v}_o = (d_a - d_b)V_d. \tag{12.12}$$

Example 12.1

In a DC motor drive, the DC-bus voltage is $V_d = 350$ V. Determine the following: $\bar{v}_{com}$, $\bar{v}_{aN}$, and d_a for pole-a and similarly for pole-b, if the output voltage required is (a) $\bar{v}_0 = 300$V and (b) $\bar{v}_0 = 300$V.

Solution From Equation 12.9, $\bar{v}_{com} = \dfrac{V_d}{2} = 175$V.

 a. For $\bar{v}_0 = 300$V, from Equation 12.8, $\bar{v}_{an} = \bar{v}_o/2 = 150$V and $\bar{v}_{bn} = -\bar{v}_o/2 = -150$V. From Equation 12.10, $\bar{v}_{aN} = 325$ V and $\bar{v}_{bN} = 25$ V. From Equation 12.11, $d_a \approx 0.93$ and $d_b \approx 0.07$.

 b. For $\bar{v}_0 = -300$ V, $\bar{v}_{an} = \bar{v}_o/2 = -150$ V and $\bar{v}_{bn} = -\bar{v}_o/2 = 150$ V. Therefore from Equation 12.10, $\bar{v}_{aN} = 25$ V and $\bar{v}_{bN} = 325$ V. From Equation 12.11, $d_a \approx 0.07$ and $d_b \approx 0.93$.

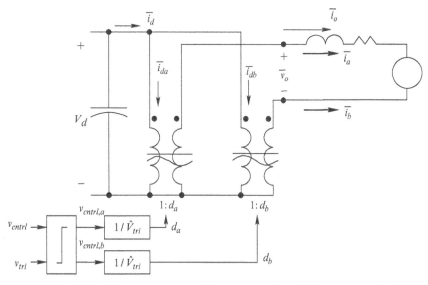

FIGURE 12.10 Switching-cycle-averaged representation of the converter for DC drives.

The switching-cycle-averaged representation of the two power poles, along with the pulse-width modulator, in a block diagram form is shown in Figure 12.10.

In each power pole of Figure 12.10, the switching-cycle-averaged DC-side current is related to its output current by the pole duty ratio,

$$\bar{i}_{da} = d_a \bar{i}_a \quad \text{and} \quad \bar{i}_{db} = d_b \bar{i}_b . \tag{12.13}$$

By Kirchhoff's current law, the total switching-cycle-averaged DC-side current is

$$\bar{i}_d = \bar{i}_{da} + \bar{i}_{db} = d_a \bar{i}_a + d_b \bar{i}_b . \tag{12.14}$$

Recognizing the directions with which the currents i_a and i_b are defined,

$$\bar{i}_a(t) = -\bar{i}_b(t) = \bar{i}_o(t). \tag{12.15}$$

Thus, substituting currents from Equation 12.14 into Equation 12.15,

$$\bar{i}_d = (d_a - d_b)\bar{i}_o. \tag{12.16}$$

Example 12.2
In the DC motor drive of Example 12.1, the output current into the motor is $\bar{i}_o = 15\,\text{A}$. Calculate the power delivered from the DC-bus and show that it is equal to the power delivered to the motor (assuming the converter to be lossless), if $\bar{v}_0 = 300\,\text{V}$.

Solution   Using the values for d_a and d_b from part (a) of Example 12.1, and $\bar{i}_o = 15\,\text{A}$ from Equation 12.16, $\bar{i}_d(t) = 12.9\,\text{A}$ and therefore the power delivered by the DC-bus is $P_d = 4.515\,\text{kW}$. Power delivered by the converter to the motor is $P_o = \bar{v}_o \bar{i}_o = 4.5\,\text{kW}$, which is equal to the input power (neglecting the round-off errors).

Using Equations 12.4 and 12.11, the control voltages for the two poles are as follows:

$$v_{cntrl,a} = \frac{\hat{V}_{tri}}{2} + \frac{\hat{V}_{tri}}{2}\left(\frac{\bar{v}_o}{V_d}\right) \quad \text{and} \quad v_{cntrl,b} = \frac{\hat{V}_{tri}}{2} - \frac{\hat{V}_{tri}}{2}\left(\frac{\bar{v}_o}{V_d}\right). \tag{12.17}$$

In Equation 12.17, defining the second term in the two control voltages above as one-half the control voltage, that is,

$$\frac{v_{cntrl}}{2} = \frac{\hat{V}_{tri}}{2}\left(\frac{\bar{v}_o}{V_d}\right). \tag{12.18}$$

Equation 12.18 simplifies as follows:

$$\bar{v}_o = \underbrace{\left(\frac{V_d}{\hat{V}_{tri}}\right)}_{K_{PWM}} v_{cntrl}, \tag{12.19}$$

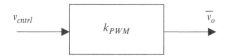

FIGURE 12.11 Gain of the converter for DC drives.

where $(V_d / \hat{V}_{tri})$ is the converter gain k_{PWM}, from the feedback control signal to the switching-cycle-averaged voltage output, as shown in Figure 12.11 in a block diagram form.

12.3.1 Switching Waveforms in a Converter for DC Motor Drives

We will look further into the switching details of the converter in Figure 12.9a. To produce a positive output voltage, the control voltages are shown in Figure 12.12. Only one-half of the time period, $T_s / 2$, needs to be considered due to symmetry.

The pole output voltages v_{aN} and v_{bN} have the same waveform as the switching signals except for their amplitude. The output voltage v_o waveform shows that the effective switching frequency at the output is twice the original. That is, within the time period of the switching frequency f_s with which the converter devices are switching, there are two complete cycles of repetition. Therefore, the harmonics in the output are at $(2f_s)$ and at its multiples. If the switching frequency is selected sufficiently large, the motor inductance may be enough to keep the ripple in the output current within an acceptable range without the need for an external inductor in series.

Next, we will look at the currents associated with this converter, repeated in Figure 12.13. The pole currents $i_a = i_o$ and $i_b = -i_o$. The DC-side current $i_d = i_{da} + i_{db}$. The waveforms for these currents are shown by means of Example 12.3.

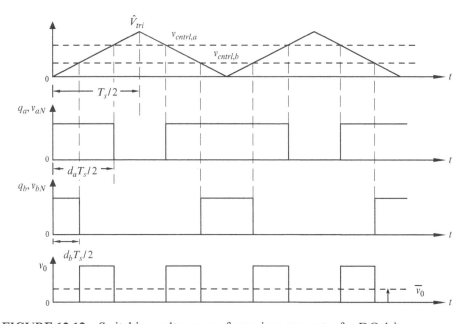

FIGURE 12.12 Switching voltage waveforms in a converter for DC drive.

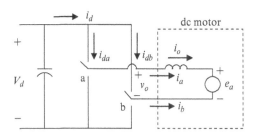

FIGURE 12.13   Currents defined in the converter for DC motor drives.

Example 12.3

In the DC motor drive of Figure 12.13, assume the operating conditions are as follows: $V_d = 350\,\text{V}$, $e_a = 236\,\text{V}(dc)$, and $\bar{i}_o = 4\,\text{A}$. The switching frequency f_s is 20 kHz. Assume that the series resistance R_a associated with the motor is $0.5\,\Omega$. Calculate the series inductance L_a necessary to keep the peak-peak ripple in the output current to be $1.0\,\text{A}$ at this operating condition. Assume that $\hat{V}_{tri} = 1\,\text{V}$. Plot v_o, $\bar{v}_o$, i_o, and i_d.

Solution   As seen from Figure 12.12, the output voltage v_o is a pulsating waveform that consists of a DC switching-cycle-averaged $\bar{v}_o$ plus a ripple component $v_{o,ripple}$, which contains sub-components that are at very high frequencies (the multiples of $2f_s$):

$$v_o = \bar{v}_o + v_{o,\,ripple}. \tag{12.20}$$

Therefore, the resulting current i_o consists of a switching-cycle-averaged DC component $\bar{i}_o$ and a ripple component $i_{o,ripple}$:

$$i_o = \bar{i}_o + i_{o,\,ripple}. \tag{12.21}$$

For a given v_o, we can calculate the output current by means of superposition by considering the circuit at DC and the ripple frequency (the multiples of $2f_s$), as shown in Figures 12.14a and 12.14b, respectively. In the DC circuit, the series inductance has no effect and hence is omitted from Figure 12.14a. In the ripple-frequency circuit of Figure 12.14b, the back-emf e_a, that is DC, is suppressed along with the series resistance R_a, which generally is negligible compared to the inductive reactance of L_a at very frequencies associated with the ripple.

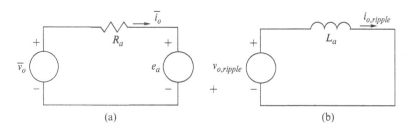

FIGURE 12.14   Superposition of DC and ripple-frequency variables.

From the circuit of Figure 12.14a,

$$\bar{v}_o = e_a + R_a\bar{i}_o = 238\,\text{V}. \tag{12.22}$$

The switching waveforms are shown in Figure 12.15, which is based on Figure 12.12, where the details are shown for the first half-cycle. The output voltage v_o pulsates between 0 and $V_d = 350\,\text{V}$, where from Equation 12.11, $d_a = 0.84$, and $d_b = 0.16$. At $f_s = 20\,\text{kHz}$, $T_s = 50\,\mu\text{s}$. Using Equations 12.20 and 12.22, the ripple voltage waveform is also shown in Figure 12.15, where during $\dfrac{d_a - d_b}{2}T_s(=17.0\,\mu\text{s})$, the ripple voltage in the circuit of Figure 12.14b is 112 V. Therefore, during this time interval, the peak-to-peak ripple ΔI_{p-p} in the inductor current can be related to the ripple voltage as follows:

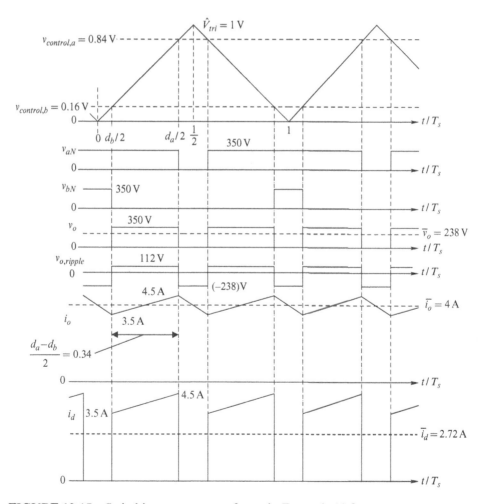

FIGURE 12.15 Switching current waveforms in Example 12.3.

$$L_a \frac{\Delta I_{p-p}}{(d_a - d_b)T_s / 2} = 112 \, \text{V}. \tag{12.23}$$

Substituting the values in the equation above with $\Delta I_{p-p} = 1 \, \text{A}$, $L_a = 1.9 \, \text{mH}$. As shown in Figure 12.15, the output current increases linearly during $(d_a - d_b)T_s / 2$, and its waveform is symmetric around the switching-cycle-averaged value; that is, it crosses the switching-cycle-averaged value at the midpoint of this interval. The ripple waveform in other intervals can be found by symmetry. The DC-side current i_d flows only during $(d_a - d_b)T_s / 2$ interval; otherwise, it is zero, as shown in Figure 12.15. Averaging over $T_s / 2$, the switching-cycle-averaged DC-side current $\bar{i}_d = 2.72 \, \text{A}$.

12.4 SYNTHESIS OF LOW-FREQUENCY AC

The principle of synthesizing a DC voltage for DC motor drives can be extended for synthesizing low-frequency AC voltages, so long as the frequency f_1 of the AC being synthesized is two or three orders of magnitude smaller than the switching frequency f_s. This is the case in most AC motor drives and UPS applications where f_1 is at 60 Hz (or is of the order of 60 Hz) and the switching frequency is a few tens of a kilohertz. The control voltage, which is compared with a triangular waveform voltage to generate switching signals, varies slowly at the frequency f_1 of the AC voltage being synthesized.

Therefore, with $f_1 << f_s$, during a switching-frequency time period $T_s(= 1/f_s)$, the control voltage can be considered pseudo-DC, and the analysis and synthesis for the converter for DC drives applies. Figure 12.16 shows how the switching power-pole output voltage can be synthesized so, on switching-cycle-averaged, it varies as shown at the low frequency f_1, where at any instant "under the microscope" shows the switching signal waveform with the duty ratio that depends on the switching-cycle-averaged voltage being synthesized. The limit on switching-cycle-averaged power-pole voltage is between 0 and V_d, as in the case of converters for DC drives.

The switching-cycle-averaged representation of the switching power-pole in Figure 12.5a is, as shown earlier in Figure 12.5b, represented by an ideal transformer with a controllable turns ratio. The harmonics in the output of the power pole in a general form were shown earlier by Figure 12.8b. In the following sections, two

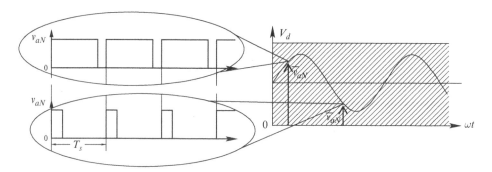

FIGURE 12.16 Waveforms of a switching power-pole to synthesize low-frequency AC.

switching power-poles are used to synthesize single-phase AC voltage for single-phase UPS and for interfacing with a single-phase supply voltage, and three switching power-poles are used to synthesize three-phase AC for motor drives, UPS, and utility-related applications.

12.5 SINGLE-PHASE INVERTERS

The load-side converter of 1-phase UPS, or for interfacing with the single-phase utility grid, is similar in power topology to that in DC motor drives, as shown in Figure 12.17a. The switching-cycle-averaged representation is shown in Figure 12.17b. It consists of two switching power-poles where, as shown, the inductance of the low-pass filter establishes the current ports of the two power poles.

The switching-cycle-averaged voltages being synthesized are shown in Figure 12.18, which in this application are sinusoidal at the line frequency f_1,

$$\bar{v}_o = \widehat{V}_o \sin \omega_1 t. \tag{12.24}$$

Similar to DC motor drives, the common-mode voltage is

$$\bar{v}_{com} = \frac{V_d}{2}, \tag{12.25}$$

and the pole output voltages with respect to a hypothetical neutral "n" are

$$\bar{v}_{an} = \frac{\bar{v}_o}{2} \quad \text{and} \quad \bar{v}_{bn} = -\frac{\bar{v}_o}{2}. \tag{12.26}$$

Therefore, as shown in Figure 12.18, the switching-cycle-averaged voltages are as follows:

$$\bar{v}_{aN} = \frac{V_d}{2} + \frac{\bar{v}_o}{2} \quad \text{and} \quad \bar{v}_{bN} = \frac{V_d}{2} - \frac{\bar{v}_o}{2}. \tag{12.27}$$

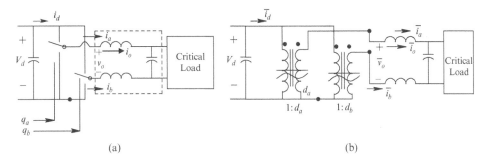

(a)   (b)

FIGURE 12.17 Single-phase uninterruptible power supply.

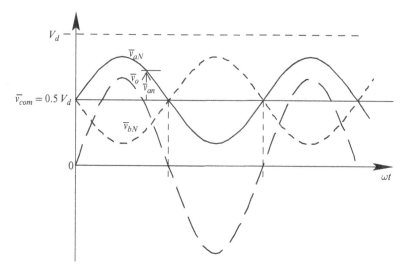

FIGURE 12.18   Switching-cycle-averaged voltages in a single-phase UPS.

On the DC-side, in order to calculate the switching-cycle-averaged current drawn from the DC source, we will assume that the switching-cycle-averaged AC-side current is sinusoidal and lagging behind the output AC voltage $\bar{v}_o(t) = \hat{V}_o \sin \omega_1 t$ by an angle ϕ_1, as shown in Figure 12.19:

$$\bar{i}_o(t) = \hat{I}_o \sin(\omega_1 t - \phi_1). \tag{12.28}$$

Assuming the ripple in the output current to be negligible, the average output power equals the product of the switching-cycle-averaged output voltage $\bar{v}_o$ and the switching-cycle-averaged output current $\bar{i}_o$,

$$P_o = \bar{v}_o \bar{i}_o. \tag{12.29}$$

Assuming the converter to be lossless, the switching-cycle-averaged input current can be calculated by equating the input power to the average power:

$$
\begin{aligned}
\bar{i}_d &= \frac{\bar{v}_o \bar{i}_o}{V_d} = \frac{\hat{V}_0 \hat{I}_o}{V_d} \sin \omega_1 t \times \sin(\omega_1 t - \phi_1) \\
&= \underbrace{0.5 \frac{\hat{V}_o}{V_d} \hat{I}_o \cos \phi_1}_{I_d} - \underbrace{0.5 \frac{\hat{V}_o}{V_d} \hat{I}_o \cos(2\omega_1 t - \phi_1)}_{i_{d2}(t)},
\end{aligned}
\tag{12.30}
$$

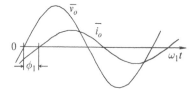

FIGURE 12.19   Output voltage and current.

which shows that the switching-cycle-averaged current drawn from the DC-bus has a DC component I_d that is responsible for the average power transfer to the AC side of the converter, and a second harmonic component i_{d2} (at twice the frequency of the AC output), which is undesirable. The DC-link storage in a 1-phase inverter must be sized to accommodate the flow of this large AC current at twice the output frequency, similar to that in PFCs, discussed in Chapter 6. Of course, we should not forget that the above discussion is in terms of switching-cycle-averaged representation of the switching power-poles. Therefore, the DC-link capacitor must also accommodate the flow of switching-frequency ripple in i_d, discussed below. The pulsating current ripple in $i_d(t)$ can be bypassed from being supplied by the batteries by placing a high-quality capacitor with a very low equivalent series inductance in close physical proximity to the converter switches.

12.5.1 Switching Waveforms Associated with a Single-Phase Inverter

The switching waveforms in a single-phase converter are shown by means of an example below.

Example 12.4

In a single-phase UPS, shown in Figure 12.17a, the parameters and the operating conditions are as follows: $V_d = 200\,\text{V}$, $f_1 = 60\,\text{Hz}$, $\overline{v}_o = 160 \sin \omega_1 t$ volts, and the switching frequency $f_s = 40\,\text{kHz}$. At the positive peak of the voltage waveform to be synthesized, calculate and plot the switching waveforms for one cycle of the switching frequency.

At the positive peak, the switching-cycle-averaged voltage to be synthesized is $\overline{v}_o = 160\,\text{V}$. Therefore, using the equations for the DC drive converters, from Equation 12.11, $d_a = 0.9$ and $d_b = 0.1$, where $T_s = 25\,\mu\text{s}$. Assuming $\hat{V}_{tri} = 1\,\text{V}$, from Equation 12.4, $V_{cntrl,a} = 0.9\,\text{V}$, and $V_{cntrl,b} = 0.1\,\text{V}$. The resulting voltage waveforms are shown in Figure 12.20.

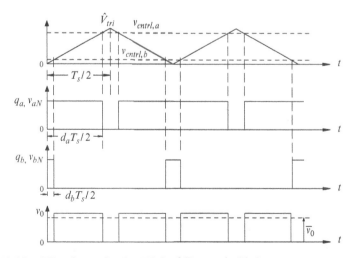

FIGURE 12.20 Waveforms in the UPS of Example 12.4.

As shown in Figure 12.20, the output voltage waveform pulsates at the switching frequency, and a low-pass filter is necessary to remove the output voltage harmonics, which were discussed earlier in a generic manner for each switching power-pole and expressed by Equation 12.5.

12.5.2   Simulation and Hardware Prototyping

The simulation of a single-phase inverter is demonstrated by means of an example:

Example 12.5

A single-phase inverter is connected to an *RL* load, $R = 8\,\Omega$, and $L = 1\,mH$. The DC voltage $V_{dc} = 200$ V. The desired output voltage is 120 V RMS at 60 Hz. Simulate this converter using LTspice.

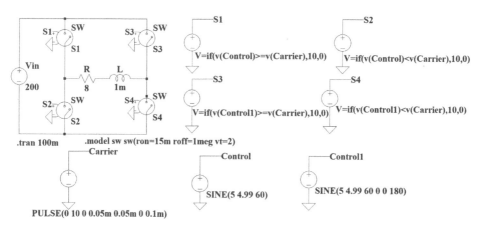

FIGURE 12.21   LTspice model.

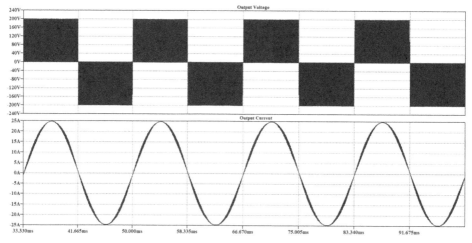

FIGURE 12.22   LTspice simulation results.

Solution The simulation file used in this example is available on the accompanying website. The LTspice model is shown in Figure 12.21, and the steady-state waveforms from the simulation of this converter are shown in Figure 12.22.

The same control algorithm used in Example 12.5 is implemented in Workbench, as shown in Figure 12.23, to generate a sinusoidal voltage from DC using the Sciamble lab kit.

In the hardware, the available DC-bus voltage is $V_{dc} = 15\,\text{V}$, and this is used to generate a 10.6 V RMS, 60 Hz output voltage, as shown in Figure 12.24. The switching frequency is chosen to be $f_s = 20\,kHz$. The step-by-step procedure for re-creating the above hardware implementation is presented in [1].

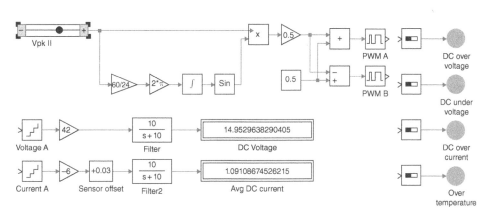

FIGURE 12.23 Workbench model.

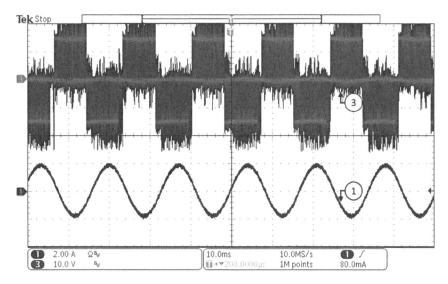

FIGURE 12.24 Workbench hardware results: (1) Output current, and (3) Output voltage.

12.6 THREE-PHASE INVERTERS

Converters for three-phase outputs consist of three power poles, as shown in Figure 12.25a. The application may be motor drives, three-phase UPS, or a three-phase utility system. The switching-cycle-averaged representation is shown in Figure 12.25b.

In Figure 12.25, $\bar{v}_{an}$, $\bar{v}_{bn}$ and $\bar{v}_{cn}$ are the desired balanced three-phase switching-cycle-averaged voltages to be synthesized: $\bar{v}_{an} = \hat{V}_{ph}\sin(\omega_1 t)$, $\bar{v}_{bn} = \hat{V}_{ph}\sin(\omega_1 t - 120°)$ and $\bar{v}_{cn} = \hat{V}_{ph}\sin(\omega_1 t - 240°)$. In series with these, common-mode voltages are added such that,

$$\bar{v}_{aN} = \bar{v}_{com} + \bar{v}_{an}\quad \bar{v}_{bN} = \bar{v}_{com} + \bar{v}_{bn}\quad \bar{v}_{cN} = \bar{v}_{com} + \bar{v}_{cn}. \qquad (12.31)$$

These voltages are shown in Figure 12.26a. The common-mode voltages do not appear across the load; only $\bar{v}_{an}$, $\bar{v}_{bn}$, and $\bar{v}_{cn}$ appear across the load with respect to the load neutral. This can be illustrated by applying the principle of superposition to the circuit in Figure 12.26a.

By "suppressing" $\bar{v}_{an}$, $\bar{v}_{bn}$, and $\bar{v}_{cn}$, only equal common-mode voltages are present in each phase, as shown in Figure 12.26b. If the current in one phase is i, then it will be the same in the other two phases. By Kirchhoff's current law at the load neutral, $3i = 0$ and hence $i = 0$, and therefore, the common-mode voltages do not appear across the load phases.

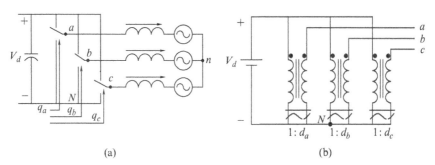

FIGURE 12.25 Three-phase converter.

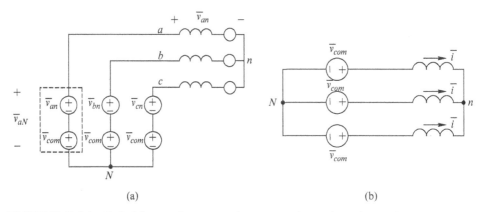

FIGURE 12.26 Switching-cycle-averaged output voltages in a three-phase converter.

To obtain the switching-cycle-averaged currents drawn from the voltage port of each switching power-pole, we will assume the currents drawn by the motor load in Figure 12.25b to be sinusoidal but lagging with respect to the switching-cycle-averaged voltages in each phase by an angle ϕ_1, where $\bar{v}_{an}(t) = \hat{V}_{ph}\sin\omega_1 t$ and so on:

$$\bar{i}_a(t) = \hat{I}\sin(\omega_1 t - \phi_1),\ \bar{i}_b(t) = \hat{I}\sin(\omega_1 t - \phi_1 - 120°),$$
$$\bar{i}_c(t) = \hat{I}\sin(\omega_1 t - \phi_1 - 240°). \tag{12.32}$$

Assuming that the ripple in the output currents is negligibly small, the average power output of the converter can be written as

$$P_o = \bar{v}_{aN}\bar{i}_a + \bar{v}_{bN}\bar{i}_b + \bar{v}_{cN}\bar{i}_c. \tag{12.33}$$

Equating the average output power to the power input from the DC-bus and assuming the converter to be lossless,

$$\bar{i}_d(t)V_d = \bar{v}_{aN}\bar{i}_a + \bar{v}_{bN}\bar{i}_b + \bar{v}_{cN}\bar{i}_c. \tag{12.34}$$

Making use of Equation 12.31 into Equation 12.34,

$$\bar{i}_d(t)V_d = \bar{v}_{com}(\bar{i}_a + \bar{i}_b + \bar{i}_c) + \bar{v}_{an}\bar{i}_a + \bar{v}_{bn}\bar{i}_b + \bar{v}_{cn}\bar{i}_c. \tag{12.35}$$

By Kirchhoff's current law at the load neutral, the sum of all three-phase currents within brackets in Equation 12.35 is zero,

$$\bar{i}_a + \bar{i}_b + \bar{i}_c = 0. \tag{12.36}$$

Therefore, from Equation 12.35,

$$\bar{i}_d(t) = \frac{1}{V_d}(\bar{v}_{an}\bar{i}_a + \bar{v}_{bn}\bar{i}_b + \bar{v}_{cn}\bar{i}_c). \tag{12.37}$$

In Equation 12.37, the sum of the products of phase voltages and currents is the three-phase power being supplied to the motor. Substituting for phase voltages and currents in Equation 12.37,

$$\bar{i}_d(t) = \frac{\hat{V}_{ph}\hat{I}}{V_d}\begin{bmatrix}\sin(\omega_1 t)\sin(\omega_1 t - \phi_1) + \sin(\omega_1 t - 120°)\sin(\omega_1 t - \phi_1 - 120°)\\ + \sin(\omega_1 t - 240°)\sin(\omega_1 t - \phi_1 - 240°)\end{bmatrix}, \tag{12.38}$$

which simplifies to a DC current, as it should, in a three-phase circuit:

$$\bar{i}_d(t) = I_d = \frac{3}{2}\frac{\hat{V}_{ph}\hat{I}}{V_d}\cos\phi_1. \tag{12.39}$$

In three-phase converters, there are two methods of synthesizing sinusoidal output voltages, both of which we will investigate:

1. Sine-PWM
2. SV-PWM (Space Vector PWM)

12.6.1 Sine-PWM

In Sine-PWM (similar to converters for DC motor drives and 1-phase UPS), the switching-cycle-averaged output of power poles, $\bar{v}_{aN}$, $\bar{v}_{bN}$, and $\bar{v}_{cN}$, has a constant DC common-mode voltage $\bar{v}_{com} = \dfrac{V_d}{2}$, similar to that in DC motor drives and single-phase UPS $\bar{v}_{an}$, $\bar{v}_{bn}$, and $\bar{v}_{cn}$ can vary sinusoidally as shown in Figure 12.27:

$$\bar{v}_{aN} = \frac{V_d}{2} + \bar{v}_{an} \qquad \bar{v}_{bN} = \frac{V_d}{2} + \bar{v}_{bn} \qquad \bar{v}_{cN} = \frac{V_d}{2} + \bar{v}_{cn}. \tag{12.40}$$

In Figure 12.27, using Equation 12.3, the plots of $\bar{v}_{aN}$, $\bar{v}_{bN}$, and $\bar{v}_{cN}$, each divided by V_d, are also the plots of d_a, d_b, and d_c within the limits of 0 and 1:

$$d_a = \frac{1}{2} + \frac{\bar{v}_{an}}{V_d} \qquad d_b = \frac{1}{2} + \frac{\bar{v}_{bn}}{V_d} \qquad d_c = \frac{1}{2} + \frac{\bar{v}_{cn}}{V_d}. \tag{12.41}$$

These power-pole duty ratios define the turns ratio in the ideal transformer representation of Figure 12.25b. As can be seen from Figure 12.27, at the limit, $\bar{v}_{an}$ can become a maximum of $\dfrac{V_d}{2}$ and hence the maximum allowable value of the phase-voltage peak is

$$(\widehat{V}_{ph})_{max} = \frac{V_d}{2}. \tag{12.42}$$

Therefore, using the properties of three-phase circuits where the line-line voltage magnitude is $\sqrt{3}$ times the phase-voltage magnitude, the maximum amplitude of the line-line voltage in sine-PWM is limited to

$$(\widehat{V}_{LL})_{max} = \sqrt{3}(\widehat{V}_{ph})_{max} = \frac{\sqrt{3}}{2}V_d \approx 0.867 V_d. \tag{12.43}$$

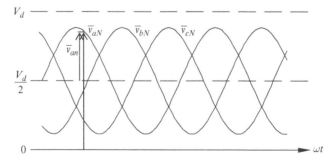

FIGURE 12.27 Switching-cycle-averaged voltages due to sine-PWM.

12.6.1.1 Switching Waveforms in a Three-Phase Inverter with Sine-PWM

In sine-PWM, three sinusoidal control voltages equal the duty ratios, given in Equation 12.41, multiplied by $\hat{V}_{tri}$. These are compared with a triangular waveform signal to generate the switching signals. These switching waveforms for sine-PWM are shown by an example below.

Example 12.6

In a three-phase converter of Figure 12.25a, a sine-PWM is used. The parameters and the operating conditions are as follows: $V_d = 350$ V, $f_1 = 60$ Hz, $\bar{v}_{an} = 160 \cos \omega_1 t$ volts, and similarly for "b" and "c" phases, and the switching frequency $f_s = 25$ kHz. $V_{tri} = 1$ V. At $\omega_1 t = 15°$, calculate and plot the switching waveforms for one cycle of the switching frequency.

Solution At $\omega_1 t = 15°$, $\bar{v}_{an} = 154.55$ V, $\bar{v}_{bn} = -41.41$ V, and $\bar{v}_{cn} = -113.14$ V. Therefore, from Equation 12.40, $\bar{v}_{aN} = 329.55$ V, $\bar{v}_{bN} = 133.59$ V, and $\bar{v}_{cN} = 61.86$ V. From Equation 12.41, the corresponding power-pole duty ratios are $d_a = 0.942$, $d_b = 0.382$, and $d_c = 0.177$. For $\hat{V}_{tri} = 1$ V, these duty ratios also equal the control voltages in volts. The switching time period $T_s = 50 \ \mu s$. Based on this, the switching waveforms are shown in Figure 12.28.

12.6.1.2 Simulation and Hardware Prototyping

The simulation of a three-phase inverter modulated using sine-PWM is demonstrated by means of an example:

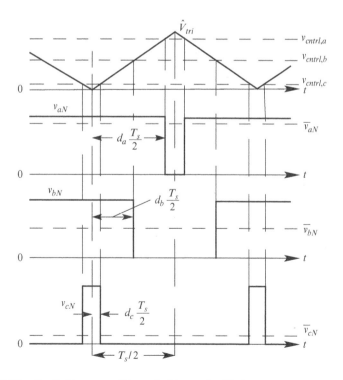

FIGURE 12.28 Switching waveforms in Example 12.6.

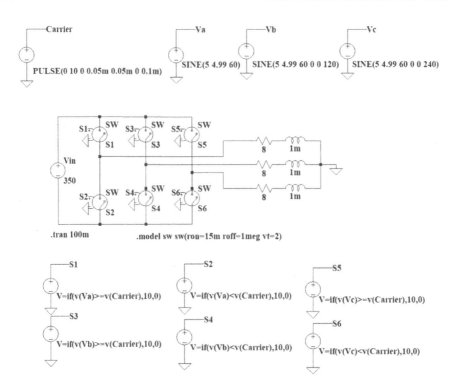

FIGURE 12.29   LTspice model.

Example 12.7

A three-phase inverter is connected to a balanced three-phase *RL* load, $R = 8\,\Omega$, and $L = 1\,mH$. The DC voltage $V_{dc} = 350\,V$. The desired output voltage is 208 V line-line RMS at 60 Hz. Simulate this converter using LTspice.

Solution   The simulation file used in this example is available on the accompanying website. The LTspice model is shown in Figure 12.29, and the steady-state waveforms from the simulation of this converter are shown in Figure 12.30.

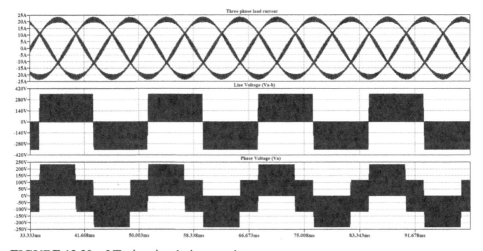

FIGURE 12.30   LTspice simulation results.

The same control algorithm used in Example 12.7 is implemented in Workbench, as shown in Figure 12.31, to generate a balanced three-phase sinusoidal voltage from DC using the Sciamble lab kit.

In the hardware, the available DC-bus voltage is $V_{dc} = 24$ V, and this is used to generate a 14.7 V RMS, 60 Hz output voltage, as shown in Figure 12.32. The switching frequency is chosen to be $f_s = 20\,kHz$. The step-by-step procedure for re-creating the above hardware implementation is presented in [2].

12.6.2 Space Vector PWM (SV-PWM)

The use of space vectors has been introduced on a physical basis such that it can be used in teaching the first course dealing with 3-phase AC machines [3]. This approach has numerous benefits compared to conventional methods of understanding AC machines.

The voltage space vector is a compact way to represent all the three-phase voltages desired at any instant by a single variable. This switching-cycle averaged space

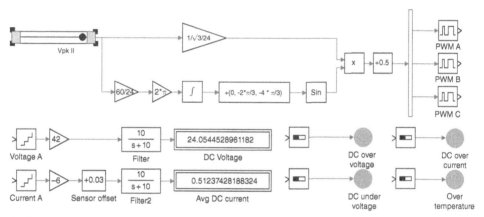

FIGURE 12.31   Workbench model.

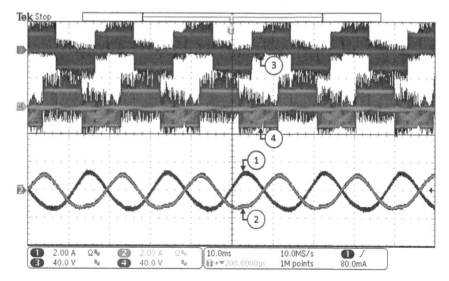

FIGURE 12.32   Workbench hardware results: (1)I_A, (2) I_B, (3) V_{AB}, and (4) V_{BC}.

vector is synthesized using space-vector PWM (SV-PWM), which fully utilizes the available DC-bus voltage and results in the AC output, which can be approximately 15% higher than that possible by using the sine-PWM approach, both in a linear range, where no lower-order harmonics appear.

Sine-PWM is limited to $(\hat{V}_{LL})_{\max} \simeq 0.867 V_d$, as given by Equation 12.43, because it synthesizes output voltages on a per-pole basis, which does not take advantage of the three-phase properties. Physically, by considering line-line voltages, it is possible to get $(\hat{V}_{LL})_{\max} = V_d$ in SV-PWM.

12.6.2.1 Definition of Space Vectors

Space vectors can be easily understood by considering a balanced three-phase load, for example, an AC machine, as shown in Figure 12.25, where the a-axis is taken as the reference axis, and the phase axes for the other two phases are $2\pi/3$ and $4\pi/3$ radians away in the counterclockwise direction.

The stator space voltage vector at any instant is defined as follows, by multiplying the stator phase voltages at that instant by their respective axes' orientation and summing them:

$$\vec{v}_s(t) = v_a(t)e^{j0} + v_b(t)e^{j2\pi/3} + v_c(t)e^{j4\pi/3}. \tag{12.44}$$

In terms of the inverter output voltages with respect to the negative DC bus "N" in Figure 12.33, and hypothetically assuming the stator neutral as a reference ground,

$$v_a = v_{aN} + v_N; v_b = v_{bN} + v_N; v_c = v_{cN} + v_N. \tag{12.45}$$

Substituting Equations 12.45 into Equation 12.44 and recognizing that

$$e^{j0} + e^{j2\pi/3} + e^{j4\pi/3} = 0, \tag{12.46}$$

the instantaneous stator voltage space vector can be written in terms of the inverter output voltages as

$$\vec{v}_s(t) = v_{aN}e^{j0} + v_{bN}e^{j2\pi/3} + v_{cN}e^{j4\pi/3}. \tag{12.47}$$

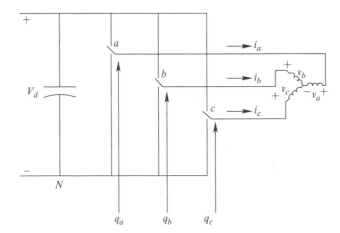

FIGURE 12.33   Inverter with a three-phase output.

A switch in an inverter pole of Figure 12.33 is in the "up" position if the pole-switching function $q = 1$ and in the "down" position if $q = 0$. In terms of the switching functions, the instantaneous voltage space vector can be written as

$$\vec{v}_s(t) = V_d(q_a e^{j0} + q_b e^{j2\pi/3} + q_c e^{j4\pi/3}). \tag{12.48}$$

With three poles, eight switch-status combinations are possible. In Equation 12.48, the instantaneous stator voltage vector $\vec{v}_s(t)$ can take on one of the following seven distinct instantaneous values where in a digital representation, phase "a" represents the least significant digit and phase "c" the most significant digit (for example, the resulting voltage vector due to the switch-status combination $\underset{(=3)}{011}$ is represented as $\vec{v}_3$):

$$\begin{aligned}
\vec{v}_s(000) &= \vec{v}_0 = 0 \\
\vec{v}_s(001) &= \vec{v}_1 = V_d e^{j0} \\
\vec{v}_s(010) &= \vec{v}_2 = V_d e^{j2\pi/3} \\
\vec{v}_s(011) &= \vec{v}_3 = V_d e^{j\pi/3} \\
\vec{v}_s(100) &= \vec{v}_4 = V_d e^{j4\pi/3} \\
\vec{v}_s(101) &= \vec{v}_5 = V_d e^{j5\pi/3} \\
\vec{v}_s(110) &= \vec{v}_6 = V_d e^{j\pi} \\
\vec{v}_s(111) &= \vec{v}_7 = 0.
\end{aligned} \tag{12.49}$$

In Equation 12.49, $\vec{v}_0$ and $\vec{v}_7$ are the zero vectors because of their zero value. The resulting instantaneous stator voltage vectors, which we will call the "basic vectors," are plotted in Figure 12.34. The basic vectors form six sectors, as shown in Figure 12.34.

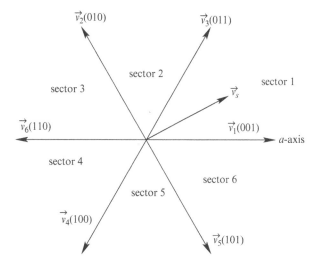

FIGURE 12.34 Basic voltage vectors ($\vec{v}_0$ and $\vec{v}_7$ are not shown).

It should be noted that the basic voltage vectors are the only true instantaneous voltages. However, the machine phase voltages that we are aiming to synthesize, for example, in Equation 12.44, are the switching-cycle-averaged voltages, and therefore, to be rigorous, they should be written as $\overline{v}_a(t)$, and so on. Therefore, the switching-cycle-averaged space vector in Equation 12.44 will have to be written as $\overline{\overrightarrow{v}}_s(t)$. This is not done in the analysis presented here simply to avoid complicating the symbols, but the intent should be clearly understood. Reiterating, as shown in Figure 12.34, the basic vectors are the truly instantaneous space vectors, which by time-weighted averaging called SV-PWM and, as discussed in the next section, allow us to synthesize the space vector $\overline{v}_s(t)$ of Figure 12.34, which is switching-cycle-averaged.

12.6.2.2 SV-PWM

The objective of the PWM control of the inverter switches is to synthesize the desired reference stator voltage space vector in an optimum manner with the following objectives:

- A constant switching frequency f_s;
- Smallest instantaneous deviation from its reference value;
- Maximum utilization of the available DC-bus voltages;
- Lowest ripple in the motor current;
- Minimum switching loss in the inverter.

The above conditions are generally met if the average voltage vector is synthesized by means of the two instantaneous basic nonzero voltage vectors that form the sector (in which the average voltage vector to be synthesized lies) and both the zero-voltage vectors, such that each transition causes the change of only one switch status to minimize the inverter switching loss.

In the following analysis, we will focus on the average voltage vector in sector 1 with the aim of generalizing the discussion to all sectors. To synthesize an average voltage vector $\overline{v}_s (= \hat{V}_s e^{j\theta_s})$ over a time period T_s, as shown in Figure 12.35, the adjoining basic vectors $\overline{v}_1$ and $\overline{v}_3$ are applied for intervals xT_s and yT_s, respectively, and the zero vectors $\overline{v}_0$ and $\overline{v}_7$ are applied for a total duration of zT_s.

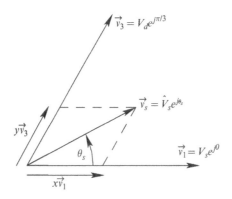

FIGURE 12.35 Voltage vector in sector 1.

In terms of the basic voltage vectors, the average voltage vector can be expressed as

$$\vec{v}_s = \frac{1}{T_s}[xT_s\vec{v}_1 + yT_s\vec{v}_3 + zT_s \cdot 0], \tag{12.50}$$

or

$$\vec{v}_s = x\vec{v}_1 + y\vec{v}_3, \tag{12.51}$$

where

$$x + y + z = 1. \tag{12.52}$$

In Equation 12.51, expressing voltage vectors in terms of their amplitude and phase angles results in

$$\hat{V}_s e^{j\theta_s} = xV_d e^{j0} + yV_d e^{j\pi/3}. \tag{12.53}$$

By equating real and imaginary terms on both sides of Equation 12.53, we can solve for x and y (in terms of the given values of V_s, θ_s, and V_d) to synthesize the desired average space vector in sector 1.

Having determined the durations for the adjoining basic vectors and the two zero vectors, the next task is to relate the above discussion to the actual poles (a, b, and c). Note in Figure 12.34 that in any sector, the adjoining basic vectors differ in one position. For example, in sector 1 with the basic vectors $\vec{v}_1(001)$ and $\vec{v}_3(011)$, only the

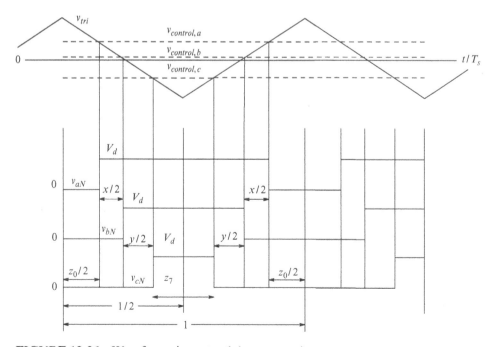

FIGURE 12.36 Waveforms in sector 1 ($z = z_0 + z_7$).

pole b differs in the switch position. For sector 1, the switching pattern in Figure 12.36 shows that pole a is in "up" position during the sum of xT_s, yT_s, and z_7T_s intervals and hence for the longest interval of the three poles. Next in the length of duration in the "up" position is pole b for the sum of yT_s and z_7T_s intervals. The smallest in the length of duration is pole c for only the z_7T_s interval. Each transition requires a change in switch state in only one of the poles, as shown in Figure 12.36. Similar switching patterns for the three poles can be generated for any other sector.

12.6.2.3 Limit on the Amplitude $\widehat{V}_s$ of the Stator Voltage Space Vector $\vec{v}_s$

First, we will establish the absolute limit on the amplitude $\widehat{V}_s$ of the average stator voltage space vector at various angles. The limit on the amplitude equals V_d (the DC-bus voltage) if the average voltage vector lies along a nonzero basic voltage vector. In between the basic vectors, the limit on the average voltage vector amplitude is that its tip can lie on the straight lines shown in Figure 12.37, forming a hexagon.

However, the maximum amplitude of the output voltage $\vec{v}_s$ should be limited to the circle within the hexagon in Figure 12.37 to prevent distortion in the resulting currents. This can be easily concluded from the fact that in a balanced sinusoidal steady state, the voltage vector $\vec{v}_s$ rotates at the synchronous speed $\omega_{syn}(=2\pi f)$, with its constant amplitude, where f is the frequency of the phase voltages. At its maximum amplitude,

$$\vec{v}_{s,\max}(t) = \widehat{V}_{s,\max} e^{j\omega_{syn}t}. \tag{12.54}$$

Therefore, the maximum value that $\widehat{V}_s$ can attain is

$$\widehat{V}_{s,\max} = V_d \cos(\frac{60°}{2}) = \frac{\sqrt{3}}{2}V_d. \tag{12.55}$$

In a balanced steady state, the peak of the phase voltages is 2/3 times the amplitude of the space vector. Therefore, from Equation 12.55, the corresponding limits on the phase voltage and the line-line voltages are as follows:

$$\widehat{V}_{phase,\max} = \frac{2}{3}\widehat{V}_{s,\max} = \frac{V_d}{\sqrt{3}}, \tag{12.56}$$

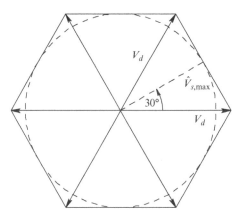

FIGURE 12.37 Limit on amplitude $\widehat{V}_s$.

and

$$V_{LL,\max}(\text{rms}) = \sqrt{3}\,\frac{\widehat{V}_{phase,\max}}{\sqrt{2}} = \frac{V_d}{\sqrt{2}} = 0.707\,V_d\,(\text{SV} - \text{PWM}). \tag{12.57}$$

The sine-PWM in the linear range, as discussed before, results in a maximum voltage

$$V_{LL,\max}(\text{rms}) = \frac{\sqrt{3}}{2\sqrt{2}}\,V_d = 0.612\,V_d \quad (\text{Sine - PWM}). \tag{12.58}$$

A comparison of Equations 12.57 and 12.58 shows that the SV-PWM discussed in this chapter better utilizes the DC-bus voltage and results in a higher limit on the available output voltage by a factor of $(2/\sqrt{3})$, or by approximately 15% higher, compared to the sine-PWM.

Example 12.8

Similar to that in Example 12.6, consider the three-phase converter of Figure 12.21a, where $V_d = 350\,\text{V}$ and $\overline{v}_{an} = 160\cos\omega_1 t$ volts, and so forth. Obtain the duty ratios d_a, using SV-PWM at $\omega_1 t = 15°$.

Solution   The voltage space vector at $\omega_1 t = 15°$ is obtained using Equation 12.47:

$$\overline{v}_s(t) = 160\cos(15°)e^{j0} + 160\cos(15° - 120°)e^{j2\pi/3} + 160\cos(15° - 240°)e^{j4\pi/3}$$
$$= 240\angle 15°$$

This is in sector 1 of the voltage vectors shown in Figure 12.34. In sector 1, The two nonzero vectors are $\overline{v}_1(001)$ and $\overline{v}_2(011)$. The ratio of the switching time period each of these vectors is applied is determined using Equation 12.53:

$$240e^{j15\pi/180} = 350\left(xe^{j0} + ye^{j\pi/3}\right)$$
$$\Rightarrow 0.6857(0.9659 + j0.2588) = (x + 0.5y) + j0.866y.$$

Solving the above equation gives $x = 0.5599$ and $y = 0.2049$. The two zero vectors $\overline{v}_0(000)$ and $\overline{v}_7(111)$ are shared equally in the remaining period:

$$z_0 = z_1 = (1 - x - y)/2 = 0.1176$$

Now the duty cycle of each phase can be obtained by summing up the duty cycles of each vector whose phase's switch is in the "up" position during that phase: $d_a = x + y + z_7 = 0.8824$, $d_b = y + z_7 = 0.3225$, and $d_c = z_7 = 0.1176$.

12.6.2.4 Simulation and Hardware Prototyping

The simulation of a three-phase inverter modulated using SV-PWM is demonstrated by means of an example:

Example 12.9

A three-phase inverter is connected to a balanced three-phase RL load, where $R = 8\,\Omega$ and $L = 1\,mH$. The DC voltage $V_{dc} = 350\,\text{V}$. The desired output voltage is 208 V line-line RMS at 60 Hz. Simulate this converter using LTspice.

Solution   The simulation file used in this example is available on the accompanying website. The LTspice model is shown in Figure 12.38, and the steady-state waveforms from the simulation of this converter are shown in Figure 12.39.

The same control algorithm used in Example 12.9 is implemented in Workbench, as shown in Figure 12.40, to generate a balanced three-phase sinusoidal voltage from DC using the Sciamble lab kit.

In the hardware, the available DC-bus voltage is $V_{dc} = 40\,V$, and this is used to generate a 17 V RMS, 60 Hz output voltage, as shown in Figure 12.41. The switching frequency is chosen to be $f_s = 20\,kHz$. The step-by-step procedure for re-creating the above hardware implementation is presented in [4].

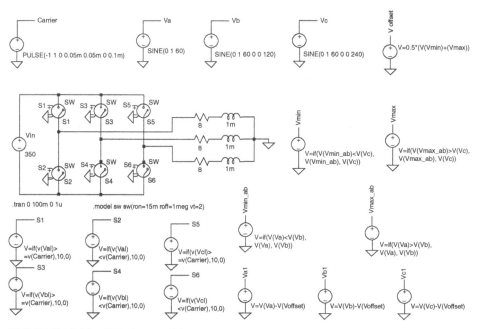

FIGURE 12.38   LTspice model.

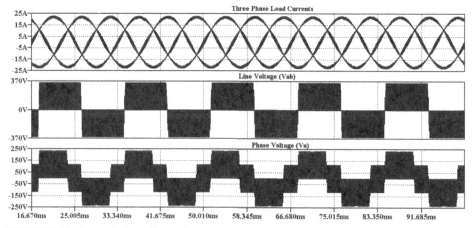

FIGURE 12.39   LTspice simulation results.

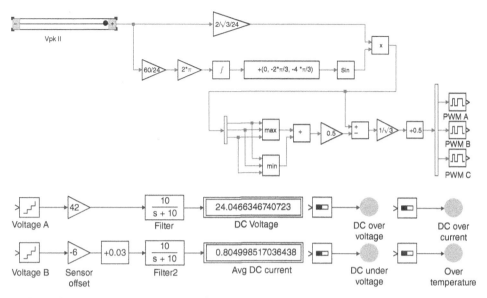

FIGURE 12.40 Workbench model.

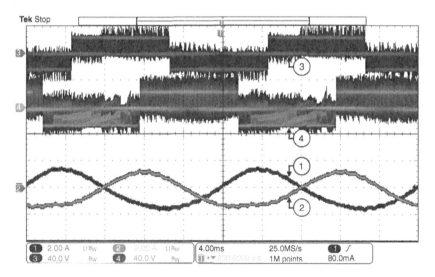

FIGURE 12.41 Workbench hardware results: (1)I_A, (2) I_B, (3) V_{AB}, and (4) V_{BC}.

12.6.3 Over-Modulation and Square-Wave (Six-step) Mode of Operation [5]

So far, in using sine-PWM and SV-PWM, it is assumed that the control voltage peak is kept equal to or less than the triangular waveform peak $\widehat{V}_{tri}$. This results in a linear range where the output phase voltages, ignoring the common-mode offset voltages that do not appear across the load, are linearly related to the control voltages. Therefore, in terms of functionality, a PWM converter is similar to a linear amplifier in the linear modulation. In addition to this linearity, the harmonics in the switched output waveform are, as shown in Figure 12.8b, at around the multiples of the switching frequency. That is, the low-order harmonics that are multiples of the low-frequency f_1 (the fundamental frequency) do not appear in the output.

However, in applications such as motor drives at higher than rated speed, it may be advantageous to get as high a voltage as possible at the fundamental frequency f_1, even if the output contains harmonics that are low-order multiples of f_1. This requires the control voltages to exceed the triangular-waveform peak by over-modulation. The output voltage-switching waveform, as a consequence, contains low-order harmonics that can be obtained by Fourier analysis. At the limit, in each switching power-pole, the switch is in "up" position for one-half the time period and "down" for the other half, as shown in Figure 12.42.

The output waveforms of the three poles are displaced by $2\pi/3$ radians with respect to each other. The resulting output waveforms are square waves, and such an operation is called square-wave or six-step mode of operation. For a given DC-bus voltage V_d, this mode of operation yields the highest possible output voltages at frequency f_1, where by Fourier analysis, at the fundamental frequency,

$$\hat{V}_{ph_1} = \frac{4}{\pi}\left(\frac{V_d}{2}\right) = 0.637 V_d \quad \text{and} \quad \hat{V}_{LL_1} = \sqrt{3}\,\hat{V}_{ph_1} = \frac{2\sqrt{3}}{\pi} V_d \simeq 1.1 V_d, \quad (12.59)$$

which shows that it is possible to get the fundamental-frequency line-line voltage peak greater than V_d, although at the expense of the low-order harmonic voltages,

$$\hat{V}_{LL_h} = \frac{\hat{V}_{LL_1}}{h} = \frac{1.1}{h} V_d \quad \text{where} \quad h = 6n \pm 1; n = 1, 2, 3, \ldots \quad (12.60)$$

12.7 MULTILEVEL INVERTERS

In high-power and high-voltage applications, it is desirable to operate with high values of voltages in order to keep the associated currents to manageable levels. This requires the DC-bus voltage V_d to be large so that it exceeds the voltage ratings of the

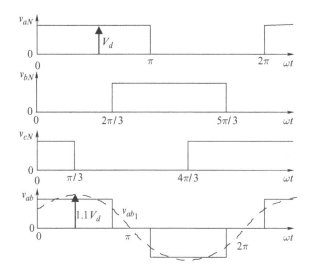

FIGURE 12.42 Square-wave (six-step) waveforms.

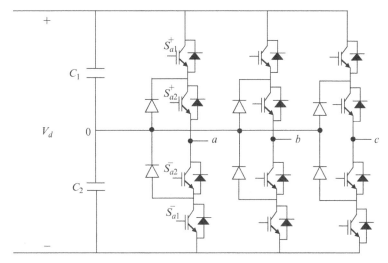

FIGURE 12.43   Three-level inverters.

transistors (of course, a safety margin has to be used), as shown in Figure 12.3a of a switching power-pole. One option in such a case is to use multiple transistors in series in order to yield, effectively, each transistor in Figure 12.3a. This is done in practice; however, great care must be taken to ensure that all the transistors in series switch in unison so that they share voltages equally.

Another option that has been used sometimes is to have a three-level arrangement, as shown in Figure 12.43, where a midpoint "o" is established by two series-connected equal capacitors as shown with equal DC voltages $V_d / 2$ [4].

We will consider the switching power-pole for phase *a*. By turning switches S_{a2}^+ and S_{a2}^- on, with respect to the midpoint, the switching-pole output voltage $v_{ao} = 0$ regardless of the current direction. Turning S_{a1}^+ and S_{a2}^+ on results in $v_{ao} = V_d / 2$. Similarly, turning S_{a1}^- and S_{a2}^- on results in $v_{ao} = -V_d / 2$. There are two major advantages of the three-level inverter: (1) transistors in Figure 12.43 need to block only one-half the DC-bus voltage, that is, $V_d / 2$, without the need to connect two transistors in series and without the associated problem of ensuring equal voltage sharing mentioned earlier, and (2) three levels ($V_d / 2, 0, -V_d / 2$) result in less switching-frequency ripple in the output for the same switching frequency, as compared to two-level inverters discussed earlier. One of the drawbacks of these inverters is the need to ensure that the midpoint remains at half the DC bus voltage.

Multilevel inverters with more than three levels have been reported in the literature with various advantages and challenges [6, 7]. A detailed discussion of this topic is presented in [8].

12.8   CONVERTERS FOR BIDIRECTIONAL POWER FLOW

In many applications, the power flow through the voltage-link structure of Figure 12.1 is bidirectional. For example, in motor drives, normally, power flows from the utility to the motor, and while slowing down, the energy stored in the inertia of machine-load combination can be recovered by operating the machine as a generator and feeding

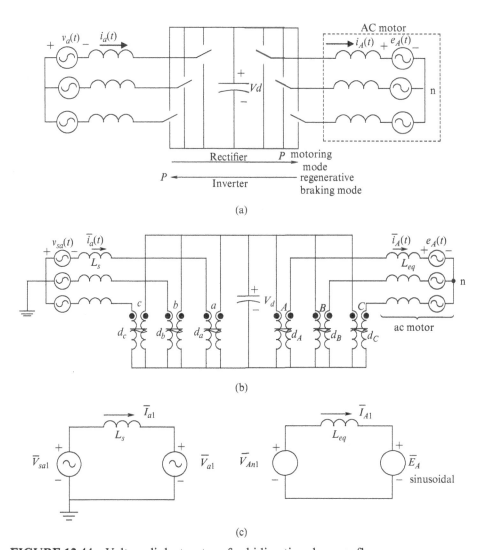

FIGURE 12.44   Voltage-link structure for bidirectional power flow.

power back into the utility grid. This can be accomplished by using three-phase converters, discussed in section 12.6, at both ends, as shown in Figure 12.44a, recognizing
that the power flow through these converters is bidirectional.

In the normal mode, the converter at the utility-end operates as a rectifier and
the converter at the machine-end as an inverter. The roles of these two are opposite
when the power flows in the reverse direction during energy recovery. A similar
arrangement can be used in connecting two AC systems by means of a HVDC transmission line.

The switching-cycle-averaged representation of these converters by means of
ideal transformers is shown in Figure 12.44b. In this simplified representation, where
losses are ignored, one side is represented by a source v_{sa}, and so forth, in series with

the internal system inductance L_s. The other side, for example, an AC machine, is represented by its steady-state equivalent circuit, the equivalent machine inductance L_{eq} in series with the back-emf e_A, and so on. Under balanced three-phase operation at both ends, the role of each converter can be analyzed by means of the per-phase equivalent circuits shown in Figure 12.44c. In these per-phase equivalent circuits, the fundamental-frequency voltages produced at the AC side by the two converters are v_{a1} and v_{An1}, and the AC-side currents at the two sides can be expressed in the phasor form as

$$\bar{I}_{a1} = \frac{\bar{V}_{sa} - \bar{V}_{a1}}{j\omega_s L_s} \tag{12.61}$$

$$\bar{I}_{A1} = \frac{\bar{V}_{An1} - \bar{E}_A}{j\omega_1 L_{eq}}, \tag{12.62}$$

where the phasors in Equation 12.61 represent voltages and current at the frequency ω_s of side 1 and the phasors in Equation 12.62 at the fundamental-frequency ω_1 synthesized at side 2.

In Figure 12.44a, for a given utility voltage, it is possible to control the current drawn from side 1 by controlling the voltage synthesized by the side 1 converter in magnitude and phase. In these circuits, if the losses are ignored, the switching-cycle-averaged power drawn from side 1 equals the power supplied to side 2. However, the reactive power at the side 1 converter can be controlled independently of the reactive power at the side 2 converter.

12.9 MATRIX CONVERTERS (DIRECT LINK SYSTEM)

To review once again, power electronics systems are categorized as voltage-link systems described so far, where a capacitor is used in parallel with two converters as an energy storage element, and current-link systems, described in Chapters 13 and 14, used in very high-power applications, where an inductor is used in series with the two converters for energy storage. There is another structure called the matrix converters, which provides a direct link between the input and the output without any intermediate energy storage element. There is a great deal of research interest at present in these converters because they avoid the intermediate energy storage element.

Such a system for a three-phase to three-phase conversion is shown in Figure 12.45, where there is a bidirectional switch from each input port to each output port. Such a bidirectional switch must be capable of blocking voltages of either polarity and conduct current in either direction. Such a bidirectional switch can be realized, for example, by two IGBTs and two diodes, as connected in Figure 12.45. When the two transistors are gated on, the current can flow in either direction, effectively representing the closed position of the bidirectional switch. When both transistors are gated off, current cannot flow in either direction, effectively representing the closed position of the bidirectional switch. A detailed discussion of this topic is presented in [8].

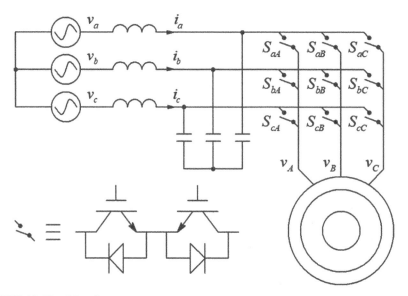

FIGURE 12.45   Matrix converter.

REFERENCES

1. "Single-Phase Inverter Lab Manual." https://sciamble.com/resources/pe-drives-lab/basic-pe/single-phase-inverter.
2. "Sine PWM Lab Manual." https://sciamble.com/resources/pe-drives-lab/basic-pe/sine-pwm.
3. N. Mohan and S. Raju, *Analysis and Control of Electric Drives: Simulations and Laboratory Implementation* (New York: John Wiley & Sons, 2020).
4. "Space Vector PWM Lab Manual." https://sciamble.com/resources/pe-drives-lab/basic-pe/svpwm.
5. N. Mohan, T.M. Undeland, and W.P. Robbins, *Power Electronics: Converters, Applications and Design*, 3rd Edition (New York: John Wiley & Sons, 2003).
6. J. Rodríguez, et al., "Multilevel Inverters: A Survey of Topologies, Controls, and Applications," *IEEE Transactions on Industrial Electronics* 49, no. 4 (August 2002): 724–738.
7. V.T. Somasekhar, K. Gopakumar, M.R. Baiju, K.K. Mohapatra, and L. Umanand, "A Multilevel Inverter System for an Induction Motor with Open-End Windings," *IEEE Transactions on Industrial Electronics* 52, no. 3 (June 2005), 824–836.
8. N. Mohan, W. Robins, T. Undeland, and S. Raju, *Power Electronics for Grid-Integration of Renewables: Analysis, Simulations and Hardware Lab* (New York: John Wiley & Sons, 2023).

PROBLEMS

Switching Power-Pole

12.1   In a switch-mode converter pole a, $V_d = 175\,\text{V}$, $\widehat{V}_{tri} = 5\,\text{V}$, and $f_s = 20\,\text{kHz}$. Calculate the values of the control signal $v_{cntrl,a}$ and the pole duty ratio d_a during which the switch is in its top position, for the following values of the average output voltage: $\bar{v}_{aN} = 125\,\text{V}$ and $\bar{v}_{aN} = 50\,\text{V}$.

12.2   In a converter pole, including the ripple in the $i_a(t)$ waveform, show that the relationship between the currents on both sides of the switching-cycle-averaged power pole is similar to that in an ideal transformer.

DC-MOTOR DRIVES

12.3 A switch-mode DC-DC converter uses a PWM-controller IC that has a triangular waveform signal at 25 kHz with $\widehat{V}_{tri} = 1.5$ V. If the input DC source voltage $V_d = 150$ V, calculate the gain k_{PWM} in Equation 12.19 of this switch-mode amplifier.

12.4 In a switch-mode DC-DC converter, $v_{cntrl} / \widehat{V}_{tri} = 0.75$ with a switching frequency $f_s = 20$ kHz and $V_d = 150$ V. Calculate and plot the ripple in the output voltage $v_o(t)$.

12.5 A switch-mode DC-DC converter is operating at a switching frequency of 20 kHz, and $V_d = 150$ V. The average current being drawn by the DC motor is 8.0 A. In the equivalent circuit of the DC motor, $E_a = 100$ V, $R_a = 0.25\,\Omega$, and $L_a = 4$ mH. (a) Plot the output current and calculate the peak-to-peak ripple, and (b) plot the current on the DC side of the converter.

12.6 In Problem 12.5, the motor goes into regenerative braking mode. The average current being supplied by the motor to the converter during braking is 7.0 A. Plot the voltage and current waveforms on both sides of this converter at that instant. Calculate the average power flow into the converter.

12.7 In Problem 12.5, calculate $\bar{i}_{da}$, $\bar{i}_{db}$, and $\bar{i}_d (= I_d)$.

12.8 Repeat Problem 12.5 if the motor is rotating in the reverse direction, with the same current draw and the same induced emf E_a value of the opposite polarity.

12.9 Repeat Problem 12.8 if the motor is braking while it has been rotating in the reverse direction. It supplies the same current and produces the same induced emf E_a value of the opposite polarity.

12.10 Repeat Problem 12.5 if a bipolar voltage switching is used in the DC-DC converter. In such a switching scheme, the two bi-positional switches are operated in such a manner that when switch a is in the top position, switch b is in its bottom position, and vice versa. The switching signal for pole a is derived by comparing the control voltage (as in Problem 12.5) with the triangular waveform.

SINGLE-PHASE INVERTERS

12.11 In a 1-phase UPS, $V_d = 350$ V, $\bar{v}_o(t) = 170\sin(2\pi \times 60t)$V, and $\bar{i}_o(t) = 10\sin(2\pi \times 60t - 30°)A$. Calculate and plot $d_a(t)$, $d_b(t)$, $\bar{v}_{aN}(t)$, $\bar{v}_{bN}(t)$, I_d, $i_{d2}(t)$, and $\bar{i}_d(t)$. Switching frequency $f_s = 20$ kHz.

12.12 In Problem 12.11, calculate $q_a(t)$ and $q_b(t)$ at $\omega t = 90°$.

THREE-PHASE INVERTERS

12.13 Plot $d_a(t)$ if the output voltage of the converter pole a is $\bar{v}_{aN}(t) = \dfrac{V_d}{2} + 0.85\dfrac{V_d}{2}\sin(\omega_1 t)$, where $\omega_1 = 2\pi \times 60$ rad / s.

12.14 In a three-phase DC-AC inverter, $V_d = 350$ V, $\widehat{V}_{tri} = 1$ V , the maximum value of the control voltage reaches 0.8 V, and $f_1 = 45$ Hz. Calculate and plot (a) the duty ratios $d_a(t)$, $d_b(t)$, $d_c(t)$, (b) the pole output voltages $\bar{v}_{aN}(t)$, $\bar{v}_{bN}(t)$, $\bar{v}_{cN}(t)$, and (c) the phase voltages $\bar{v}_{an}(t)$, $\bar{v}_{bn}(t)$, and $\bar{v}_{cn}(t)$.

12.15 In a balanced three-phase DC-AC converter, the phase a average output voltage is $\bar{v}_{an}(t) = 112.5\sin(\omega_1 t)$, where $V_d = 300$ V and $\omega_1 = 2\pi \times 45$ rad$/$s. The inductance L in each phase is 5 mH. The AC motor internal voltage in phase A can be represented as $e_a(t) = 106.14\sin(\omega_1 t - 6.6°)$V. (a) Calculate and plot $d_a(t)$, $d_b(t)$, and $d_c(t)$, and (b) sketch $\bar{i}_a(t)$ and $\bar{i}_{da}(t)$.

12.16 In Problem 12.15, calculate and plot $\bar{i}_d(t)$, which is the average DC current drawn from the DC side.

12.17 In a converter $V_d = 700$ V. To synthesize an average stator voltage vector $\vec{v}_s^a = 563.38e^{j0.44}$ V, calculate x, y, and z.

12.18 Given that $\vec{v}_s^a = 563.38e^{j0.44}$ V, calculate the phase voltage components.

SIMULATION PROBLEMS

12.19 Simulate a single-phase inverter where $V_d = 300$ V, the output voltage v_0 is 150 V (RMS) at the fundamental frequency, which is 45 Hz. The output current i_0 has an RMS value of 10 A at a lagging power factor of 0.866. The switching frequency is $f_s = 1$ kHz. The output load can be simulated by a back-emf in series with a resistance of $2.0\,\Omega$ and an inductance of 10.0 mH.
   (a) Obtain v_0 and i_0 waveforms.
   (b) By Fourier analysis, obtain v_{01}, and plot the v_{01} and i_0 waveforms.
   (c) Obtain the i_d waveform.
   (d) By Fourier analysis, obtain I_d and i_{d2}, and plot them.
   (e) Obtain the RMS value of the high-frequency ripple current in i_d.

12.20 In the single-phase inverter of Problem 12.17, represent each of the two poles by their average model.
   (a) Obtain the $\bar{v}_0$ and $\bar{i}_0$ waveforms.
   (b) Obtain the $\bar{i}_d$ waveform.
   (c) Obtain I_d and i_{d2}, and plot them.

12.21 Simulate a three-phase inverter where $V_d = 350$ V, the output voltage V_{LL} is 175 V (RMS) at the fundamental frequency, which is 45 Hz. The output current has an RMS value of 10 A at a lagging power factor of 0.866. The switching frequency is $f_s = 1$ kHz. The per-phase output load can be simulated by a back-emf in series with a resistance of $2.0\,\Omega$ and an inductance of 10.0 mH.
   (a) Obtain the waveforms for v_{an} (with respect to load-neutral), i_a and i_d.
   (b) Obtain v_{an1} by means of Fourier analysis of the v_{an} waveform.
   (c) Using the results of part (b), obtain the ripple component v_{ripple} waveform in the output voltage.

12.22 In the three-phase inverter of Problem 12.19, represent each of the three poles by their average model.
   (a) Observe the waveforms for $\bar{v}_{an}$, $\bar{i}_a$, and $\bar{i}_d$.
   (b) Append the output current waveforms of the switching model of Problem 12.19.

13

THYRISTOR CONVERTERS

13.1 INTRODUCTION

Historically, thyristor converters were used to perform tasks that are now performed by switch-mode converters, discussed in previous chapters. Thyristor converters are now typically used in utility applications at very high power levels. In this chapter, we will examine the operating principles of thyristor-based converters.

13.2 THYRISTORS (SCRs)

A thyristor is a device that can be considered a controlled diode. Like diodes, they are available in very large voltage and current ratings, making them attractive for use in applications at very high power levels.

Thyristors, shown by their symbol in Figure 13.1a, are sometimes referred to by their trade name of silicon controlled rectifiers (SCRs). These are 4-layer (p-n-p-n) devices, as shown in Figure 13.1b. When a reverse ($v_{AK} < 0$) voltage is applied, the flow of current is blocked by the junctions pn1 and pn3. When a forward ($v_{AK} > 0$) polarity

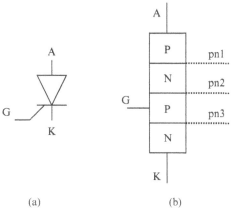

(a)   (b)

FIGURE 13.1 Thyristors.

Power Electronics A First Course: Simulations and Laboratory Implementations, Second Edition.
Ned Mohan and Siddharth Raju.
© 2023 John Wiley & Sons, Inc. Published 2023 by John Wiley & Sons, Inc.
Companion Website: www.wiley.com/go/mohan/powerelectronics2e

voltage is applied, and the gate terminal is open, the flow of current is blocked by the junction pn2, and the thyristor is considered to be in a forward-blocking state. In this forward-blocking state, applying a small positive voltage to the gate with respect to the cathode for a short interval supplies a pulse of gate current i_G that latches the thyristor in its on state, and subsequently, the gate-current pulse can be removed.

13.2.1 Primitive Thyristor Rectifier Circuits

The behavior of thyristors is illustrated by means of a simple circuit with a resistive load in Figure 13.2a. At $\omega t = 0$, the positive half-cycle of the input voltage begins, beyond which a forward voltage appears across the thyristor (anode A is positive with respect to cathode K), and if the thyristor were a diode, a current would begin to flow in this circuit. This instant we will refer to as *the instant of natural conduction*. With the thyristor in a forward-blocking state, the start of conduction can be controlled (delayed) with respect to the instant of natural conduction, which is $\omega t = 0$ in this circuit, by a delay angle α at which instant the gate-current pulse is applied. Once in the conducting state, the thyristor behaves like a diode with a very small voltage drop of the order of 1 to 2 volts across it (we will idealize it as zero), and the load voltage v_d equals v_s in Figure 13.2b, where v_d is indicated by the darker waveform. The current equals v_d / R, as shown in Figure 13.2b.

The current declines to zero at $\omega t = \pi$, and since it cannot reverse through the thyristor, it stays zero during the negative half-cycle of the voltage waveform, as shown in Figure 13.2b. The current through the thyristor stays zero until the gate pulse is applied in the next cycle of the voltage waveform. The average value V_d of the load voltage is indicated by the dotted line in Figure 13.2b. This average value can be calculated analytically from the V_d waveform in Figure 13.2b as

$$V_d = \frac{1}{2\pi} \int_{\alpha}^{\pi} \hat{V}_s \sin\omega t \cdot d(\omega t) = \frac{\hat{V}_s}{2\pi}(1 + \cos\alpha), \tag{13.1}$$

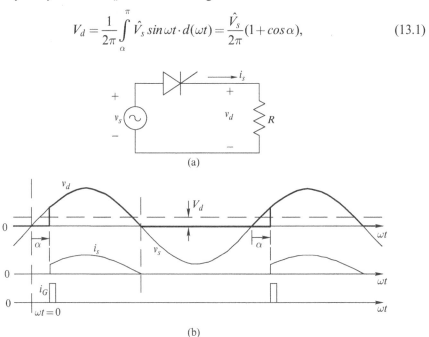

(a)

(b)

FIGURE 13.2 A simple thyristor circuit with a resistive load.

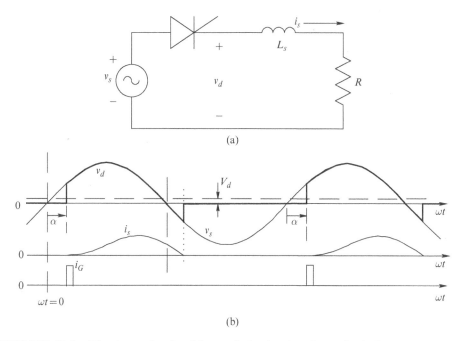

FIGURE 13.3 Thyristor circuit with a resistive load and a series inductance.

where $\hat{V}_s$ is the peak of the input AC voltage. From Equation 13.1, it is clear that we can control the average load voltage by controlling the delay angle α; this was not possible in the diode-rectifier circuits of Chapter 5.

To consider the influence of inductance in series, consider the primitive circuit of Figure 13.3a. The associated waveforms in Figure 13.3b show that due to the inductor, the current builds up slowly and comes to zero some time in the negative half-cycle of the input voltage. The current through the thyristor cannot reverse and remains zero for the remainder of the input voltage cycle. This principle can be extended to practical circuits discussed below.

13.3 SINGLE-PHASE, PHASE-CONTROLLED THYRISTOR CONVERTERS

Figure 13.4a shows a commonly used full-bridge phase-controlled converter for controlled rectification of the single-phase utility voltage. To understand the operating principle, it is redrawn in Figure 13.4b, where the AC-side inductance L_s is ignored, and the DC-side load is represented as drawing a constant current I_d. The waveforms are shown in Figure 13.5.

Thyristors (1, 2) and (3, 4) are treated as two pairs, where each thyristor pair is supplied gate pulses delayed by an angle α with respect to the instant of natural conduction at $\omega t = 0°$ for thyristors (1, 2) and at $\omega t = 180°$ for 3 and 4, as shown in Figure 13.5.

In the positive half-cycle of the input voltage, thyristors 1 and 2 are forward-blocking until they are gated at $\omega t = \alpha$ when they immediately begin to conduct I_d

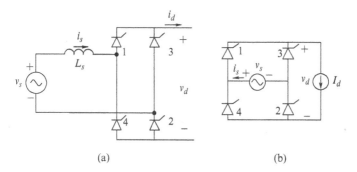

FIGURE 13.4 Full-bridge, single-phase thyristor converter.

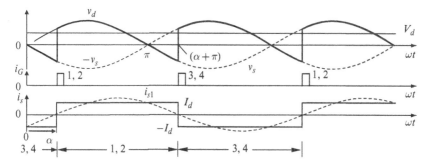

FIGURE 13.5 Single-phase thyristor converter waveforms.

(because L_s is assumed to be zero), and thyristors 3 and 4 become reverse-blocking. In this state,

$$v_d(t) = v_s(t) \text{ and } i_s(t) = I_d, \quad \alpha < \omega t \leq \alpha + \pi. \tag{13.2}$$

These relationships hold true until $\alpha + \pi$ in the negative half-cycle of the input voltage when thyristors 3 and 4 are gated and begin conducting I_d. In this state,

$$v_d(t) = -v_s(t) \quad \text{and} \quad i_s(t) = -I_d \quad \alpha + \pi < \omega t \leq \alpha + 2\pi, \tag{13.3}$$

which holds true for one half-cycle until the next half-cycle begins with the gating of thyristors 1 and 2.

The average value V_d of the voltage across the DC side of the converter can be obtained by averaging the $v_d(t)$ waveform in Figure 13.5 over only one half-cycle (by symmetry) during $\alpha < \omega t \leq \alpha + \pi$:

$$V_d = \frac{1}{\pi} \int\limits_{\alpha + \pi}^{\alpha} \hat{V}_s \sin \omega t \cdot d(\omega t) = \frac{2}{\pi} \hat{V}_s \cos \alpha. \tag{13.4}$$

On the AC side, the input current i_s waveform is shifted by an angle α with respect to the input voltage, as shown in Figure 13.5, and the fundamental-frequency component $i_{s1}(t)$ has a peak value of

$$\hat{I}_{s1} = \frac{4}{\pi} I_d. \tag{13.5}$$

In terms of voltage and current peak values, the power drawn from the AC side is

$$P = \frac{1}{2} \hat{V}_s \hat{I}_{s1} \cos \alpha. \tag{13.6}$$

Assuming no power loss in the thyristor converter, the input power equals the power to the DC side of the converter. Using Equations 13.4 and 13.5, we can reconfirm the following relationship:

$$P = \frac{1}{2} \hat{V}_s \hat{I}_{s1} \cos \alpha = V_d I_d. \tag{13.7}$$

In this converter, the current is unidirectional, but the DC-side voltage can be controlled and can be of either polarity. Therefore, the power flow can be controlled by the delay angle α; increasing it toward 90° reduces the average DC-side voltage V_d while simultaneously shifting the input current $i_s(t)$ waveform farther away with respect to the input voltage waveform. The DC voltage as a function of α is plotted in Figure 13.6a, and the corresponding power direction is shown in Figure 13.6b. The waveforms in Figure 13.5 show that for the delay angle α in a range from 0° to 90°, V_d has a positive value, as also plotted in Figures 13.6a and 13.6b, and the converter operates as a rectifier, with power flowing from the AC side to the DC side.

Delaying the gating pulse such that α is greater than 90° in Figure 13.5 makes V_d negative (also confirmed by Equation 13.4), and the converter operates as an inverter, as illustrated in Example 13.1.

Example 13.1
Draw the waveforms for the full-bridge thyristor converter of Figure 13.4b if it's operating in an inverter mode with the delay angle α equal to 150°.

Solution   Since α now equals 150°, in comparison to Figure 13.5, the i_s waveform is shifted by 150° with respect to v_s waveform as shown in Figure 13.7.

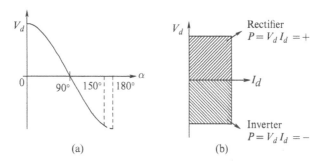

(a)   (b)

FIGURE 13.6   Effect of the delay angle α.

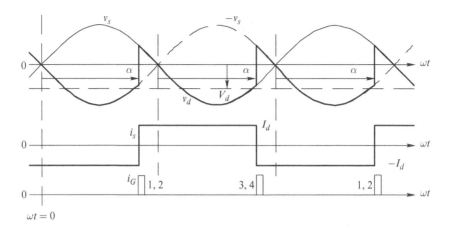

FIGURE 13.7   Single-phase thyristor converter in an inverter mode with $\alpha = 150°$.

In the inverter mode, as shown in Figure 13.6b, power flows from the DC side to the AC side. In practical circuits with inductance in series on the AC side, the upper limit on α in the inverter mode is less than $180°$, for example, $160°$, as shown in Figure 13.6a, to avoid a phenomenon known as the commutation failure, where the current fails to commutate fully from the conducting thyristor pair to the next pair, prior to the instant beyond which the conducting pair keeps on conducting for another half-cycle. This commutation-failure phenomenon is described in greater detail in [1].

13.3.1   Current Harmonics and Reactive Power Requirement

As can be seen from the AC-side current waveform i_s in Figure 13.5, the current is rectangular, and the fundamental-frequency waveform is drawn dotted, whose amplitude $\hat{I}_{s1}$ is given in Equation 13.5. From Fourier analysis, the harmonics h of i_s can be expressed in terms of the fundamental frequency component as

$$\hat{I}_{sh} = \frac{\hat{I}_{s1}}{h} \quad \text{(where } h \text{ takes on odd values 3,5,7, etc.)} \tag{13.8}$$

As can be observed from Figure 13.5, i_{s1} is displaced with respect to v_s by the delay angle α. Therefore, the reactive power drawn by the converter is

$$Q = \frac{1}{2}\hat{V}_s \hat{I}_{s1} \sin\alpha. \tag{13.9}$$

13.3.2   Effect of L_s on Current Commutation

Previously, our assumption was that the AC-side inductance L_s equals zero. Now consider the circuit in Figure 13.8a. In this case, the input current takes a finite amount of time to reverse its direction through the AC-side inductance as the current "commutates" from one thyristor pair to the next.

From basic principles, we know that changing the current through the inductor L_s in the circuit of Figure 13.8a requires a finite amount of volt-seconds. The DC side is still represented by a DC current I_d. The waveforms are shown in Figure 13.8b, where thyristors 3 and 4 are conducting prior to $\omega t = \alpha$, and $i_s = -I_d$.

At $\omega t = \alpha$, thyristors 1 and 2, which have been forward blocking, are gated, and hence they immediately begin to conduct. However, the current through them doesn't jump instantaneously as in the case of $L_s = 0$ where i_s instantaneously changed from $(-I_d)$ to $(+I_d)$. With a finite L_s, during a short interval called the commutation interval u, all thyristors conduct, thus resulting in $v_d = 0$ and applying v_s across L_s. To correspond to Figure 13.8b, where waveforms are plotted with respect to ωt, we will calculate the commutation voltage integral in volt-radians rather than in volt-seconds. The volt-radians needed to change the inductor current from $(-I_d)$ to $(+I_d)$ can be calculated by integrating the inductor voltage $v_L (= L_s \cdot di_s / dt)$ from α to $(\alpha + u)$, as follows:

$$\int_{\alpha}^{\alpha+u} v_L d(\omega t) = L_s \int_{\alpha}^{\alpha+u} \frac{di_s}{dt} d(\omega t) = \omega L_s \int_{-I_d}^{I_d} \frac{di_s}{dt} dt = \omega L_s \int_{-I_d}^{I_d} di_s = 2\omega L_s I_d. \quad (13.10)$$

The above volt-radians are "lost" from the integral of the DC-side voltage waveform in Figure 13.8b every half-cycle, as shown by the shaded area A_u in Figure 13.8b. Therefore, dividing the volt-radians in Equation 13.10 by the π radians each half-cycle, the voltage drop in the DC-side voltage is

$$\Delta V_d = \frac{2}{\pi} \omega L_s I_d. \quad (13.11)$$

This voltage is lost from the DC-side average voltage in the presence of L_s. Therefore, the average voltage is smaller than that in Equation 13.4:

$$V_d = \frac{2}{\pi} \hat{V}_s \cos \alpha - \frac{2}{\pi} \omega L_s I_d. \quad (13.12)$$

We should note that the voltage drop in the presence of L_s doesn't mean a power loss in L_s; it simply means a reduction in the voltage available on the DC side.

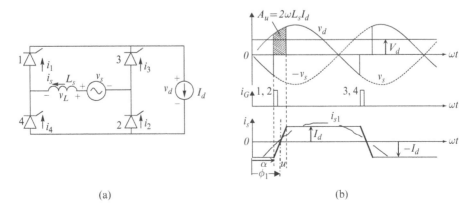

(a)   (b)

FIGURE 13.8 Effect of L_s on current commutation.

Example 13.2

In a single-phase thyristor converter of Figure 13.8a including L_s, derive the expression for the commutation angle u, and (b) calculate it if $V_s(\mathrm{rms}) = 120\,\mathrm{V}$, $L_s = 5\,\mathrm{mH}$, $I_d = 10\,\mathrm{A}$, and $\alpha = 30°$. Frequency $f = 60\,\mathrm{Hz}$.

Solution

a. From Figure 13.8a, in Equation 13.10 during the commutation interval, $v_L = v_s$. Therefore, substituting v_s for v_L in Equation 13.10,

$$\int_{\alpha}^{\alpha+u} \hat{V}_s \sin\omega t \cdot d(\omega t) = \hat{V}_s[\cos\alpha - \cos(\alpha + u)] = 2\omega L_s I_d,$$

or

$$\cos(\alpha + u) = \cos\alpha - \frac{2\omega L_s I_d}{\hat{V}_s}. \tag{13.13}$$

b. Substituting the given values in Equation 13.13, where $\hat{V}_s = \sqrt{2}\,V_s(\mathrm{rms})$, the commutation angle $u = 19.92°$.

Assuming a linear increase/decrease in thyristor currents in Figure 13.8, the fundamental frequency component i_{s1} shown dotted in Figure 13.8b lags the voltage v_s by an angle ϕ_1:

$$\phi_1 \simeq \alpha + \frac{u}{2}, \tag{13.14}$$

where the approximately-equal sign is due to the assumption of linear increase/decrease in thyristor currents. Therefore, the reactive power drawn by the converter is

$$Q \simeq \frac{1}{2}\hat{V}_s\hat{I}_{s1}\sin(\alpha + \frac{u}{2}). \tag{13.15}$$

13.4 THREE-PHASE, FULL-BRIDGE THYRISTOR CONVERTERS

Three-phase full-bridge converters use six thyristors, as shown in Figure 13.9a. A simplified converter for initial analysis is shown in Figure 13.9b, where the AC-side inductance L_s is assumed zero, thyristors are divided into a top group and a bottom group, similar to the three-phase diode rectifiers, and the DC side is represented by a current source I_d.

The converter waveforms, where the delay angle α (measured with respect to the instant at which phase voltage waveforms cross each other) is zero, are similar to those in diode rectifiers, discussed in Chapter 5 (see Figures 5.11 and 5.12). The average DC voltage is as calculated in Equation 5.33, where $\hat{V}_{LL}$ is the peak value of the AC input voltage:

$$V_{do} = \frac{1}{\pi/3} \int_{-\pi/6}^{\pi/6} \hat{V}_{LL} \cos\omega t \cdot d(\omega t) = \frac{3}{\pi}\hat{V}_{LL}. \tag{13.16}$$

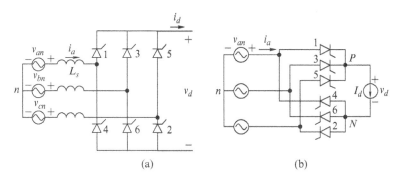

FIGURE 13.9 Three-phase full-bridge thyristor converter.

Delaying the gate pulses to the thyristors by an angle α measured with respect to their instants of natural conductions, the waveforms are shown in Figure 13.10 where $L_s = 0$.

In the DC-side output voltage waveforms, the area A_α corresponds to "volt-radians loss" due to delaying the gate pulses by α every $\pi/3$ radian. Assuming the time-origin as shown in Figure 13.10 at the instant at which v_{an} and v_{cn} waveforms cross, the line-line voltage v_{ac} waveform can be expressed as $\hat{V}_{LL} \sin \omega t$. Therefore from Figure 13.10, the drop ΔV_α in the average DC-side voltage can be calculated as

$$\Delta V_\alpha = \frac{1}{\pi/3} \underbrace{\int_0^\alpha \hat{V}_{LL} \sin \omega t \cdot d(\omega t)}_{A_\alpha} = \frac{3}{\pi} \hat{V}_{LL}(1 - \cos \alpha). \qquad (13.17a)$$

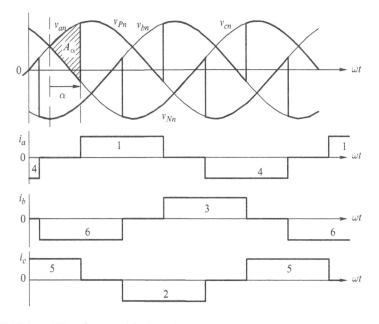

FIGURE 13.10 Waveforms with $L_s = 0$.

Using Equations 13.16 and 13.17a, with a finite delay angle α, the DC-side voltage is

$$V_{d\alpha} = \frac{3}{\pi}\hat{V}_{LL}\cos\alpha = V_{do}\cos\alpha. \tag{13.17b}$$

The phase currents in Figure 13.9a can be expressed as

$$i_a = i_1 - i_4 \quad i_b = i_3 - i_6 \quad i_c = i_5 - i_2, \tag{13.18}$$

where i_1, i_2, and so on, are the currents in the forward direction through the thyristors. Currents through all three phases are shown in Figure 13.10. These waveforms show that each thyristor conducts for 120° during a fundamental-frequency cycle, and the effect of the delay angle α is to shift current waveforms by this angle to the right, causing them to lag their input voltages by this angle.

Similar to single-phase converters, three-phase thyristor converters go into the inverter mode with the delay angle α exceeding 90°. In the inverter mode, the upper limit on α is less than 180° to avoid commutation failure, just like in single-phase converters. Further details on three-phase thyristor converters can be found in [1].

Example 13.3
Three-phase thyristor converter of Figure 13.9b is operating in its inverter mode with $\alpha = 150°$. Draw waveforms similar to Figure 13.10 for this operating condition.

Solution   These waveforms for $\alpha = 150°$ in the inverter mode are shown in Figure 13.11.

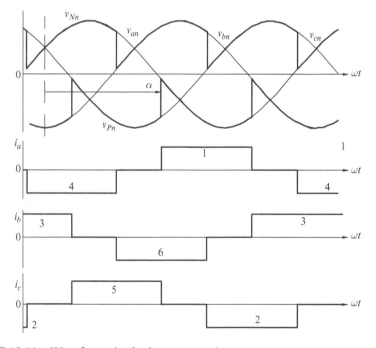

FIGURE 13.11   Waveforms in the inverter mode.

13.4.1 Current Harmonics and Reactive Power Requirement

As can be seen from the AC-side current i_a, for example, in Figure 13.10, the current has a rectangular waveform, and the fundamental-frequency component, from Fourier analysis, has an amplitude $\hat{I}_{a1}$:

$$\hat{I}_{a1} = \frac{\sqrt{12}}{\pi} I_d. \tag{13.19}$$

The harmonics h of i_a can be expressed in terms of the fundamental frequency component as

$$\hat{I}_{ah} = \frac{\hat{I}_{a1}}{h} \quad \text{(where } h = 6n \pm 1 \text{ and } n = 1, 2, 3, \text{ etc.).} \tag{13.20}$$

In Figure 13.10, i_{a1} is displaced with respect to v_{an} by the delay angle α. Therefore, the reactive power drawn by the three-phase converter is

$$Q = \frac{3}{2} \hat{V}_a \hat{I}_{a1} \sin \alpha. \tag{13.21}$$

13.4.2 Effect of L_s

Unlike in the previous section, due to the presence of L_s, it takes a finite commutation interval u for the current to commutate from one thyristor to the next, for example from thyristor 5 connected to phase c to the thyristor 1 connected to phase a. The subcircuit under discussion during the commutation interval is shown in Figure 13.12a.

During the commutation interval u, from α to $\alpha + u$, both thyristors are conducting as the current I_d tries to commutate from 5 to 1, and as derived in Example 13.4, the voltage v_{Pn} is the average of the two-phase voltages:

$$v_{Pn} = \frac{v_{an} + v_{cn}}{2}, \quad \alpha < \omega t < \alpha + u, \tag{13.22}$$

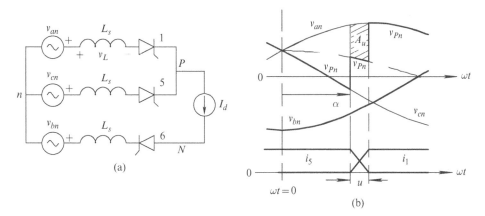

FIGURE 13.12 Commutation of current from thyristor 5 to thyristor 1.

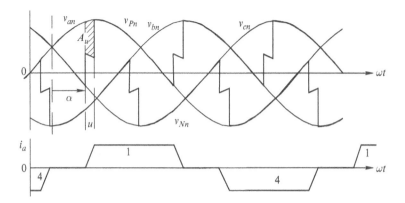

FIGURE 13.13   Waveforms with L_s.

which is shown in Figure 13.12b. As a consequence, the resulting voltage waveforms are shown in Figure 13.13, where such a commutation takes place every 60°. Considering Figure 13.12a during the commutation interval, v_{Pn} is reduced due to the voltage drop across the inductance in series with thyristor 1 to which the current is commutating from 0 to $(+I_d)$. Using the procedure in Equation 13.10 for single-phase converters, the area A_u (in volt-radians) in Figure 13.13 can be calculated as

$$A_u = \int_{\alpha}^{\alpha+u} v_L d(\omega t) = \omega L_s \int_{0}^{I_d} di_s = \omega L_s I_d. \tag{13.23}$$

Since such a commutation takes place every $\pi/3$ radians, the average DC output voltage is reduced by the area A_u divided by $\pi/3$ radians. Making use of Equation 13.23, the drop in the average DC-side voltage is

$$\Delta V_u = \frac{A_u}{\pi/3} = \frac{3}{\pi}\omega L_s I_d. \tag{13.24}$$

Therefore, the DC-side output voltage can be written as

$$V_d = V_{do} - \Delta V_\alpha - \Delta V_u. \tag{13.25}$$

Substituting results from Equations 13.16, 13.17, and 13.24 into Equation 13.25,

$$V_d = \frac{3}{\pi}\hat{V}_{LL}\cos\alpha - \frac{3}{\pi}\omega L_s I_d. \tag{13.26}$$

Example 13.4
In a three-phase thyristor converter of Figure 13.9a including L_s, (a) derive the expression for the commutation angle u, and (b) calculate it if $V_{LL}(rms) = 460\,\text{V}$, $L_s = 5\,\text{mH}$, $I_d = 20\,\text{A}$, and $\alpha = 30°$.

Solution

a. In the sub-circuit of Figure 13.12a when both thyristors are conducting, applying Kirchhoff's current law at point P,

$$i_a + i_c = I_d. \tag{13.27}$$

Assuming I_d to be constant, taking the time derivatives of both sides of Equation 13.20 results in

$$\frac{di_a}{dt} + \frac{di_c}{dt} = 0, \tag{13.28}$$

or

$$\frac{di_a}{dt} = -\frac{di_c}{dt}. \tag{13.29}$$

Therefore, in Figure 13.12a,

$$v_{Pn} = v_{an} - L_s \frac{di_a}{dt}. \tag{13.30}$$

Also,

$$v_{Pn} = v_{cn} - L_s \frac{di_c}{dt}. \tag{13.31}$$

Adding Equations 13.30 and 13.31 and making use of Equation 13.29,

$$v_{Pn} = \frac{v_{an} + v_{cn}}{2}, \qquad \alpha < \omega t < \alpha + u. \tag{13.32}$$

Substituting the expression for v_{Pn} from Equation 13.32 into Equation 13.30,

$$v_L = L_s \frac{di_a}{dt} = v_{an} - v_{Pn} = \frac{1}{2}(v_{an} - v_{cn}). \tag{13.33}$$

In Equation 13.33, $(v_{an} - v_{cn})$ is the line-line voltage, and assuming the time-origin in Figure 13.13 at the intersection of the phase-a and phase-c voltages, the voltage drop v_L in Equation 13.33 can be written as

$$v_L = L_s \frac{di_a}{dt} = \frac{1}{2}\hat{V}_{LL}\sin\omega t, \alpha < \omega t < \alpha + u \tag{13.34}$$

Using Equation 13.34 into Equation 13.23,

$$A_u = \int_{\alpha}^{\alpha+u} \frac{1}{2}\hat{V}_{LL}\sin\omega t \, d(\omega t) = \omega L_s I_d, \tag{13.35}$$

and therefore,

$$\cos(\alpha + u) = \cos\alpha - \frac{2\omega L_s I_d}{\hat{V}_{LL}}, \tag{13.36}$$

from which the commutation interval u can be calculated.

b. Substituting the given values in Equation 13.36, the commutation interval u is 11.4°.

Similar to single-phase converters, assuming a linear increase/decrease in thyristor currents in Figure 13.12b, the fundamental frequency component i_{a1} in Figure 13.13 lags the voltage v_s by

$$\phi_1 \simeq \alpha + \frac{u}{2} \tag{13.37}$$

where the approximately-equal sign is due to assuming a linear increase/decrease in thyristor currents.

13.5   CURRENT-LINK SYSTEMS

Thyristors are available in very large current and voltage ratings of several kilo-amperes and several kilovolts that can be connected in series. In addition, thyristor converters can block voltages of both polarities but conduct current only in the forward direction. This capability has led to the interface realized by thyristor-converters with a DC-current link in the middle, as shown in Figure 13.14.

Unlike in voltage-link systems, the transfer of power in current-link systems can be reversed in direction by reversing the voltage polarity of the DC link. This structure is used at very high power levels, in excess of a thousand megawatts, for example, in high-voltage DC (HVDC) transmission systems.

Thyristors in these two converters shown in Figure 13.14 are connected to allow the flow of current in the DC link by thyristors in converter 1 pointing up, and the thyristors in converter 2 connected to point downward. In Figure 13.14, subscripts 1 and 2 refer to systems 1 and 2 and R_d is the resistance of the DC link that includes the DC transmission line. Assuming each converter is a six-pulse thyristor converter, as discussed previously,

$$V_{d1} = \frac{3}{\pi}\hat{V}_{LL1}\cos\alpha_1 - \frac{3}{\pi}\omega L_{s1}I_d, \tag{13.38}$$

$$V_{d2} = \frac{3}{\pi}\hat{V}_{LL2}\cos\alpha_2 - \frac{3}{\pi}\omega L_{s2}I_d. \tag{13.39}$$

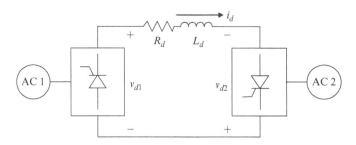

FIGURE 13.14   Block diagram of current-link systems.

By controlling the delay angles α_1 and α_2 in a range of $0°$ to $180°$ (practically, this value is limited to approximately $160°$), the average voltage and the average current in the system of Figure 13.14 can be controlled, where current flow through the DC link can be expressed as

$$I_d = \frac{V_{d1} + V_{d2}}{R_d}. \tag{13.40}$$

Since the DC-link resistance R_d is generally very small, V_{d1} and V_{d2} are very close in magnitude and opposite in value. For example, for the power flow from system 1 to system 2, V_{d2} is made negative by controlling α_2 such that converter 2 operates as an inverter and establishes the voltage of the DC link. Converter 1 is operated as a rectifier, with a positive value of V_{d1}, at a delay angle α_1 such that it controls the current in the DC link. The converse is true for these two converters if the power is to flow from system 2 to system 1.

The above discussion of current-link systems shows the operating principle behind HVDC transmission systems discussed in the next chapter, where using transformers, six-pulse thyristor converters are connected in series on the DC side and in parallel on the AC side to yield a higher effective pulse number.

REFERENCE

1. N. Mohan, T.M. Undeland, and W.P. Robbins, *Power Electronics: Converters, Applications and Design*, 3rd Edition (New York: John Wiley & Sons, 2003).

PROBLEMS

Single-Phase Thyristor Converters

In a single-phase thyristor converter, $V_s = 120\,\text{V(rms)}$ at $60\,\text{Hz}$, and $L_s = 3\,\text{mH}$. The delay angle $\alpha = 45°$. This converter is supplying $1\,\text{kW}$ of power. The DC-side current i_d can be assumed purely DC.

13.1 Calculate the commutation angle u.
13.2 Draw the waveforms for the converter variables v_s, i_s and v_d.
13.3 Assuming that the currents through the thyristors increase/decrease linearly during commutations, calculate the reactive power drawn by the converter.

Three-Phase Thyristor Converters

In a three-phase thyristor converter, $V_{LL} = 460\,\text{V(rms)}$ at $60\,\text{Hz}$, and $L_s = 5\,\text{mH}$. The delay angle $\alpha = 45°$. This converter is supplying $5\,\text{kW}$ of power. The DC-side current i_d can be assumed purely DC.

13.4 Calculate the commutation angle u.
13.4 Draw the waveforms for the converter variables: phase voltages, phase currents, v_{Pn}, v_{Nn} and v_d.

13.6 Assuming that the currents through the thyristors increase/decrease linearly during commutations, calculate the reactive power drawn by the converter.

13.7 In a three-phase thyristor converter, assume the commutation angle to be zero. Also, assume a given AC-side voltage V_{LL}(rms) and that the DC-side current I_d is kept constant in magnitude. Plot the reactive power Q drawn by the converter in terms of I_d, V_{LL}(rms), as a function of the delay angle α. Explain why it is desirable to operate the converter close to $\alpha = 0$ in the rectifier mode and close to $\alpha = 180°$ in the inverter mode.

Current-Link System

13.8 In the block diagram of Figure 13.14, for both three-phase converters, $V_{d0} = 480\,\text{kV}$. The DC-side current is $I_d = 1.1\,\text{kA}$. Converter 2 operating as an inverter establishes the DC-link voltage such that $V_{d2} = -425\,\text{kV}$. $R_d = 10.0\,\Omega$. The drop in DC voltage due to commutation overlap in each converter is $10\,\text{kV}$. In DC steady state, calculate the following angles: α_1, α_2, u_1 and u_2.

13.9 In Problem 13.8, calculate the reactive power drawn by each converter. Assume that the currents through the thyristors increase/decrease linearly during commutations.

Simulation Problems

In a single-phase thyristor converter, the input voltage is $V_s = 120\,\text{V}$ (rms) at 60 Hz frequency. The AC-side inductance is 1.2 mH, and on the DC side the load has an inductance of 20 mH in series with a resistance of 5 Ω.

(a) Obtain the v_s, v_d and i_d waveforms.
(b) Obtain the v_s and i_s waveforms.
(c) From the plots, obtain the commutation interval u.
(d) By means of Fourier analysis of i_s, calculate its harmonic components as a ratio of I_{s1}.
(e) Calculate I_s, %THD in the input current, the input displacement power factor, and the input power factor.

14

UTILITY APPLICATIONS OF POWER ELECTRONICS

14.1 INTRODUCTION

In previous chapters, various power electronic systems were described to use electricity efficiently and harness energy from renewable sources. In this chapter, all these applications, and new ones, are compiled in the context of utility applications. Such power electronics applications in utility systems are growing very rapidly, which promise to change the landscape of future power systems in terms of generation, operation, and control. The goal of this chapter is to briefly present an overview of these applications to prepare students for new challenges in the deregulated utility environment and to motivate them to take further courses in this field, for instance, the advanced-level graduate course designed at the University of Minnesota to discuss these topics in detail [1, 2].

In discussing these applications, we will observe that the power electronic converters are the same or modifications of those that we have already discussed in earlier chapters. Therefore, within the scope of this book, it will suffice to discuss these applications in terms of the block diagrams of various converters. These utility applications of power electronics can be categorized as follows:

- Distributed Generation (DG)
  - Renewable resources (wind, photovoltaic, etc.)
  - Fuel cells and micro-turbines
  - Storage-batteries, super-conducting magnetic storage, flywheels
- Power Electronic Loads—Adjustable Speed Drives
- Power Quality Solutions
  - Dual feeders
  - Uninterruptible power supplies
  - Dynamic voltage restorers

Power Electronics A First Course: Simulations and Laboratory Implementations, Second Edition.
Ned Mohan and Siddharth Raju.
© 2023 John Wiley & Sons, Inc. Published 2023 by John Wiley & Sons, Inc.
Companion Website: www.wiley.com/go/mohan/powerelectronics2e

- Transmission and Distribution (T&D): HVDC and FACTS [3]
  - High voltage DC (HVDC) and HVDC-light
  - Flexible AC transmission (FACTS)
    - Shunt compensation
    - Series compensation
    - Static phase angle control and unified power flow controllers

14.2 POWER SEMICONDUCTOR DEVICES AND THEIR CAPABILITIES

Figure 14.1a shows the commonly used symbols of power devices. The power handling capabilities and switching speeds of these devices are indicated in Figure 14.1b. All these devices allow current flow only in their forward direction (the intrinsic anti-parallel diode of MOSFETs can be explained separately). Transistors (intrinsically or by design) can block only the forward polarity voltage, whereas thyristors can block both forward and reverse polarity voltages. Diodes are uncontrolled devices that conduct current in the forward direction and block a reverse voltage. At very high power levels, integrated-gate controlled thyristors (IGCTs), which have evolved from the gate-turn-off thyristors, are used. Thyristors are semi-controlled devices that can switch-on at the desired instant in their forward-blocking state but cannot be switched off from their gate and hence rely on the circuit in which they are connected to switch them off. However, thyristors are available in very large voltage and current ratings.

These devices are available in a large range of voltage and current ratings for use in a variety of applications. Moreover, they can be connected in series and in parallel to extend this range further. As an example, IGBTs, which are commonly used in a very large range of utility-related applications, are available as a single module from a manufacturer with a rating of 600 A/6,500 V.

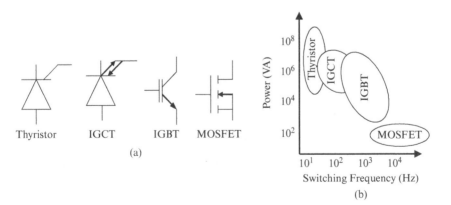

FIGURE 14.1 Power semiconductor devices.

14.3 CATEGORIZING POWER ELECTRONIC SYSTEMS

In a very broad sense, the role of power electronics in these power system applications can be categorized as follows:

14.3.1 Solid-State Switches

By connecting two thyristors in anti-parallel (back-to-back), as shown in Figure 14.2, it is possible to realize a solid-state switch that can conduct current in both directions and turn on or turn off in an AC circuit with a delay of no more than one-half the line-frequency cycle. Such switches are needed for applications such as dual feeders, shunt-compensation for injecting reactive power at a bus for voltage control, and series-compensation of transmission lines.

14.3.2 Converters as an Interface

Power electronic converters provide the needed interface between the electrical source, often the utility, and the load, as shown in Figure 14.3. The electrical source and the electrical load can, and often do, differ in frequency, voltage amplitudes, and the number of phases. The power electronics interface allows the transfer of power from the source to the load in the most energy-efficient manner, in which it is possible for the source and load to reverse roles. These interfaces can be classified as below.

14.3.2.1 Voltage-Link Systems

The semiconductor devices such as transistors of various types and diodes can only block forward-polarity voltages. These devices with only unipolar voltage-blocking capability have led to the structure with two converters, where the DC ports of these two converters are connected to each other with a parallel capacitor forming a DC

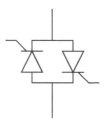

FIGURE 14.2   Back-to-back thyristors to act as a solid-state switch.

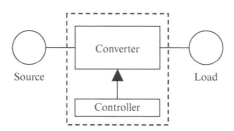

FIGURE 14.3   Power electronics interface.

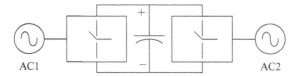

FIGURE 14.4   Block diagram of the voltage-link systems.

link, as shown in Figure 14.4, across which the voltage polarity does not reverse, thus allowing unipolar voltage-handling transistors to be used within these converters. The transfer of power can be reversed in direction by reversing the direction of currents associated with the DC-link system.

Voltage-link converters consist of switching power-poles as the building blocks, which can synthesize the desired output by means of pulse-width modulation (PWM) and have bidirectional power flow capability. Such switching power-poles can be modeled by means of an ideal transformer with a controllable turns ratio, as discussed in Chapter 12.

14.3.2.2   Current-Link Systems

Thyristors can block voltages of both polarities but conduct current only in the forward direction. This capability has led to the interface realized by thyristor-converters with a DC-current link in the middle, as shown in Figure 14.5. The transfer of power can be reversed in direction by reversing the voltage polarity of the DC link, but the currents in the link remain in the same direction. This structure is used at very high power levels, at tens of hundreds of megawatts, for example, in high-voltage DC (HVDC) transmission.

In the following sections, various utility applications and the role of power electronics in them are examined further.

14.4   DISTRIBUTED GENERATION (DG) APPLICATIONS

Distributed generation promises to change the landscape of how power systems of the future will be operated and controlled. These generators may have decentralized (local) control, in addition to a central supervisory control. There is a move away from large central power plants toward distributed generation due to environmental and economic reasons. Renewable resources such as wind and photovoltaic systems are growing in their popularity. There are proposals to place highly efficient small-scale power plants, based on fuel cells and micro-turbines, near load centers to simultaneously avoid transmission congestion and line losses. Many distributed generation systems are discussed below.

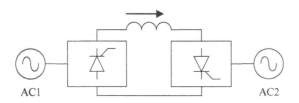

FIGURE 14.5   Block diagram of the current-link systems.

14.4.1 Wind-Electric Systems

Wind energy is an indirect manifestation of solar energy caused by uneven heating of the earth's surface by the sun. Out of all renewable energies, wind has come a long way, and still, this potential is just beginning to be realized. Figure 14.6 shows the wind potential by states in the United States, where there are several areas with good to excellent wind conditions.

Commonly used schemes for power generation in windmills require a gearing mechanism because the wind turbine rotates at very slow speeds, whereas the generator operates at high speed close to the synchronous speed, which at the 60-Hz line frequency would be 1800 rpm for a 4-pole and 900 rpm for an 8-pole machine. Therefore, the nacelle contains a gearing mechanism that boosts the turbine speed to drive the generator at a higher speed; the need for a gearing mechanism is one of the inherent drawbacks of such schemes. There are proposals to use direct-drive (without gears) permanent magnet machines in very large sizes; however, in practice, most windmills use gearing. This sub-section describes various types of wind-generation schemes.

14.4.1.1 Induction *Generators*, Directly Connected to the Grid

As shown in Figure 14.7a, this is the simplest scheme, where a wind-turbine-driven squirrel-cage induction generator is directly connected to the grid through a back-to-back connected thyristor-pair for soft start. Therefore, it is the least expensive and uses a rugged squirrel-cage rotor induction machine.

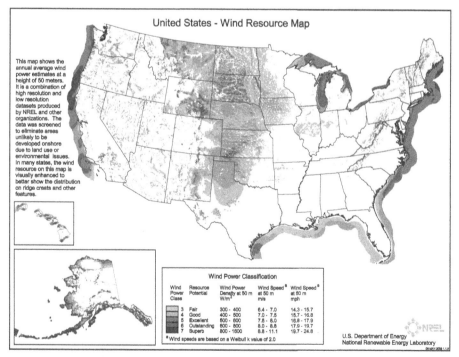

FIGURE 14.6 Wind-resource map of the United States. This information has been reprinted from the National Renewable Energy Laboratory's Dynamic Maps, GIS Data, & Analysis Tools webpage (http://www.nrel.gov/gis/wind.html), map "US Wind Resource Map." Accessed August 4, 2011.

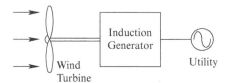

FIGURE 14.7A   Induction generator directly connected to the grid.

For the induction machine to operate in its generator mode, the rotor speed must be greater than the synchronous speed. The drawback of this scheme is that since the induction machine always operates very close to the synchronous speed, this scheme is not optimum at low and high wind speeds compared to the variable speed schemes described below. Another disadvantage of this scheme is that a squirrel-cage induction machine always operates at a lagging power factor (that is, it draws reactive power from the grid as an inductive load would). Therefore, a separate source, for example, shunt-connected capacitors, is often needed to supply the reactive power to overcome the lagging power factor operation of the induction machine.

14.4.1.2   Doubly-Fed, Wound-Rotor Induction Generators

The scheme in Figure 14.7b utilizes a wound-rotor induction machine where the stator is directly connected to the utility supply, and the rotor is injected with desired currents through a power-electronics interface. Typically, four-fifths of the power flows directly from the stator to the grid, and only one-fifth of the power flows through the power electronics in the rotor circuit. The drawback of this scheme is that it uses a wound-rotor induction machine where the currents to the three-phase wound rotor are supplied through slip-rings and brushes, which require maintenance. In spite of the fact that power electronics is expensive, and since it is rated only one-fifth of the system rating, the overall cost is not much higher than the previous scheme. However, there are several distinct advantages over the previous scheme, as described below.

The scheme using a doubly fed wound-rotor induction machine can typically operate in a range of $\pm 30\%$ around the synchronous speed, and hence it is able to capture more power at lower and higher wind speeds compared to the previous scheme because it can operate at above, as well as below, the synchronous speed. It can also supply

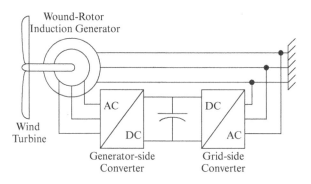

FIGURE 14.7B   Doubly-fed, wound-rotor induction generator.

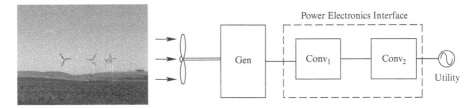

FIGURE 14.7C Power electronics connected generator.

reactive power, whereas, in the previous scheme, the squirrel-cage induction machine only absorbs reactive power. Therefore, the scheme using a doubly fed wound-rotor induction machine is quite popular in windmills being installed in the United States.

14.4.1.3 Power Electronics Connected Generator

In the third scheme shown in Figure 14.7c, a squirrel-cage induction generator or a permanent-magnet generator is connected to the grid through a power electronics interface. This interface consists of two converters. The converter at the generator-end supplies the reactive power excitation needed if it is an induction generator. Its frequency of operation is controlled to be optimal for the prevailing wind speed. The converter at the line-end is capable of absorbing or supplying reactive power in a continuous manner. This is the most flexible arrangement using a rugged squirrel-cage machine or a high-efficiency permanent-magnet generator, which can operate in a very wide wind-speed range and is the likely contender for the future arrangements as the cost of the power electronics interface, which must handle the entire power output of the system, is continuing to go down.

14.4.2 Photovoltaic (PV) Systems

Photovoltaic systems are the ultimate in distributed generation and have even greater potential than wind-electric systems. In PV systems, the PV arrays (typically four of them connected in series) provide a voltage of 52 to 90 V DC, which the power electronic system, such as that shown in Figure 14.8, converts to 120 V/60 Hz sinusoidal voltage suitable for interfacing with the single-phase utility. Such PV modules can be connected in parallel for higher output capacity. Larger arrays are interfaced with a three-phase utility grid.

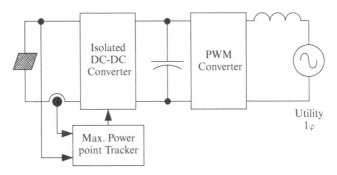

FIGURE 14.8 Photovoltaic systems.

FIGURE 14.9 A rooftop PV system. Source: www.NREL.gov

Figure 14.9 shows a 127 kW rooftop PV system, one of the largest roof-integrated federal systems in the United States.

14.4.3 Fuel Cell Systems

Lately, there has been a great deal of interest in and effort being devoted to fuel cell systems. The reason has to do with their efficiency, which can be as high as 60%. The input to the fuel cell can be hydrogen, natural gas, or gasoline, and the output is a DC voltage, as shown in Figure 14.10 [4]. The need for power electronics converters to interface with the utility is the same in the fuel-cell systems as that for the photovoltaic systems.

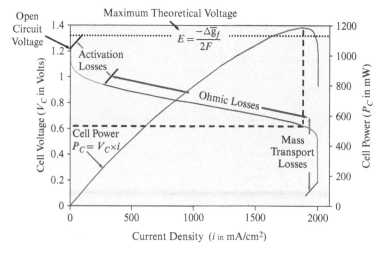

FIGURE 14.10 Fuel cell V-I relationship and cell power. Source: U.S Department of Energy / Public Domain.

14.4.4 Micro-Turbines

These use highly efficient aircraft engines, for example, to produce peak power when it is needed, with natural gas as the input fuel. To improve efficiency, the turbine speed should be allowed to vary based on loading. This will cause the frequency of the generated output to vary, requiring a power-electronic interface like that in adjustable-speed drives.

14.4.5 Energy Storage Systems

Although not a primary source of energy, storage plants offer the benefit of load-leveling and peak-shaving in power systems because of the diurnal nature of electricity usage. Energy is stored, usually at night when the load demand is low, and supplied back during the peak-load periods.

It is possible to store energy in lead-acid batteries (other exotic high-temperature batteries are being developed), in superconducting magnetic energy storage (SMES) coils, and in the inertia of flywheels. All these systems need a power electronic interface, where the interface for the flywheel storage is shown in Figure 14.11.

14.5 POWER ELECTRONIC LOADS

As discussed in Chapter 1, power electronics is playing a significant role in energy conservation, for example, as users discover the benefits of reduced energy consumption and better process control by operating electric drives at adjustable speeds. An adjustable-speed drive (ASD) is shown in Figure 14.12 in a block diagram form.

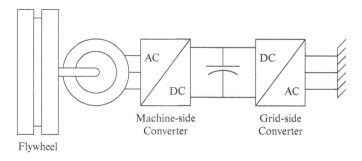

FIGURE 14.11 Flywheel storage system.

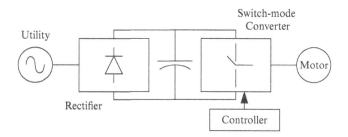

FIGURE 14.12 Adjustable-speed drive.

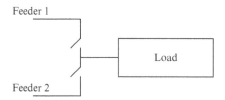

FIGURE 14.13 Dual-feeders.

Power electronic loads of this type often use an interface with the utility that results in distorted line currents. These currents result in distorted voltage waveforms, affecting the neighboring loads. However, it is possible to design the utility interface (often called the power-factor-corrected front-end) that allows sinusoidal currents to be drawn from the grid. With the proliferation of power electronic loads, standards are being enforced that limit the amount of distortion in currents drawn.

14.6 POWER QUALITY SOLUTIONS

Poor quality of power can imply any of the following: distorted voltage waveforms, unbalances, swells and sags in voltage and power outages, and so on. This problem is exacerbated in a deregulated environment where utilities are forced to operate at marginal profits, resulting in inadequate maintenance of equipment. In this section, we will also see that power electronics can solve many of the power quality problems.

14.6.1 Dual Feeders

The continuity of service can be enhanced by dual power feeders to the load, where one acts as a backup to the other that is supplying the load, as shown in Figure 14.13.

Using back-to-back connected thyristors, acting as a solid-state switch, it is possible to switch the load quickly, without interruption, from the main feeder to the backup feeder and back.

14.6.2 Uninterruptible Power Supplies

Power outages, even for a few cycles, can be very disruptive to critical loads. To provide immunity from such outages, power-electronics-based uninterruptible power supplies (UPS), shown in Figure 14.14, can be used, where the energy storage can be by means of batteries, SMES (superconducting magnetic energy storage), flywheels, or ultra-capacitors.

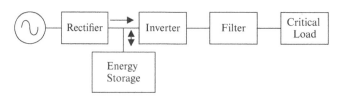

FIGURE 14.14 Uninterruptible power supplies.

14.6.3 Dynamic Voltage Restorers

Dynamic voltage restorers (DVR), shown in Figure 14.15, can compensate for voltage sags or swells by injecting a voltage v_{inj} in series with that supplied by the utility.

14.7 Transmission and Distribution (T&D) Applications

In the past, utilities operated as monopolies, where users had little or no option in selecting whom they could purchase power from. The recent trend, which seems irreversible in spite of recent setbacks, is to deregulate where utilities must compete to sell power on an open market, and the customers have a choice of selecting their power provider. In such an environment, utilities that were vertically integrated are now forced to split into generation companies that produce power and transmission-line operators that maintain the transmission and distribution network for a fee.

In this deregulated environment, it is highly desirable to have the capability to dictate the flow of power on designated power lines, avoiding overloading of transmission lines and excessive power losses in them. In this section, we will look at some such options.

14.7.1 High-Voltage DC (HVDC) Transmission

Direct current (DC) transmission represents the ultimate flexibility, isolating two interconnected AC systems from the requirement of operating in synchronism or even at the same frequency. High-voltage DC (HVDC) transmission systems using thyristor-based current-link converters have now been in operation for several decades. Lately, systems at somewhat lower voltages using IGBT-based voltage-link converters have been installed. Both of these types of systems are discussed below.

14.7.1.1 Thyristor-Based Current-Link HVDC Transmission Systems

Figure 14.16 shows the block diagram of an HVDC transmission system, where power is transmitted over DC lines at high voltages in excess of 500 kV. First, the voltages in AC system 1 at the sending end are stepped up by means of a transformer. These voltages are rectified into DC by means of a thyristor-based converter, where the AC line voltages provide the commutation of current within the converter, such that AC currents drawn from system 1 turn into DC current on the other side of the converter. These currents are transmitted over the DC line, where additional series inductance is added to ensure that the DC current is smooth and free of ripple as much as possible. At the receiving end, there are thyristor-based converters, which convert the DC current into AC currents and inject them into the AC system 2, through a step-down transformer. The roles of the sending and the receiving ends can be reversed by reversing the voltage polarity in the DC system.

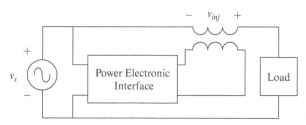

FIGURE 14.15 Dynamic voltage restorers.

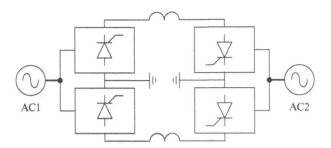

FIGURE 14.16 HVDC voltage-link system block diagram (transformers are not shown).

For example, a current-link HVDC transmission system in the western part of the United States transmits 3,100 MW at ±500 kV over a distance of 1,361 km. At very high power levels in excess of 1,000 MW, the use of thyristors, at least for now, represents the only reasonable choice.

14.7.1.2   HVDC Transmission System Using Voltage-Link IGBT-Based Converters
At lower power levels compared to those carried by thyristor-based HVDC transmission lines, the alternative is to use a voltage-link system, whose block diagram is shown in Figure 14.17. In such a system, the direction of power flow is reversed by changing the direction of current in the DC line. As an example, a voltage-link system in the eastern part of the United States operates at ±150 kV and is rated at 330 MW.

14.7.2   Flexible AC Transmission Systems (FACTS) [3]
DC transmission systems discussed earlier are an excellent choice where a large amount of power needs to be transmitted over long distances or if the system stability is a serious issue. In existing AC transmission networks, limitations on constructing new power lines and their cost have led to other ways to increase power transmission capability without sacrificing the stability margin. These techniques may also help in directing the power flow to designated lines.
Power flow on a transmission line connecting two AC systems in Figure 14.18 is given as

$$P = \frac{E_1 E_2}{X} \sin \delta, \tag{14.1}$$

where E_1 and E_2 are the magnitudes of voltages at the two ends of the transmission line, X is the line reactance, and δ is the angle between the two bus voltages.

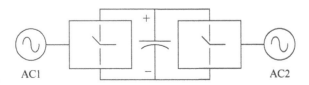

FIGURE 14.17   Block diagram HVDC transmission using a voltage-link system.

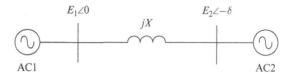

FIGURE 14.18 Power flow on a transmission line.

Equation (14.1) shows that the power flow on a transmission line depends on three quantities: (1) the voltage magnitude E, (2) the line reactance X, and (3) the power angle δ. Various devices that are based on rapidly controlling one or more of the above three quantities are discussed in the following sections:

14.7.2.1 Shunt-Connected Devices to Control the Bus Voltage Magnitude E

The reactive power compensation is very important and may even be necessary at high loadings to avoid voltage collapse. Shunt-connected devices, as shown in Figure 14.19, can draw or supply reactive power from a bus, thus controlling the bus voltage, albeit in a limited range, based on the internal system reactance.

Various forms of such devices are being used in different combinations. These include thyristor-controlled reactors (TCR), shown in Figure 14.19a, and thyristor-switched capacitors (TSC), shown in Figure 14.19b, for static var compensation (SVC). The advanced static var compensator (STATCOM) shown in Figure 14.19c can draw or supply reactive power.

Shunt-compensation devices have the following limitations for controlling the flow of active power:

1. A large amount of reactive power compensation, depending on the system's internal reactance, may be required to change the voltage magnitude. Of course, the voltage can only be changed in a limited range (utilities try to maintain bus voltages at their nominal values), which has a limited effect on the power transfer given by Equation (14.1).
2. Most transmission systems consist of parallel paths or loops. Therefore, changing the voltage magnitude at a given bus changes the loading of all the lines connected to that bus, and there is no way to dictate the desired change of power flow on a given line.

14.7.2.2 Series-Connected Devices to Control the Effective Series Reactance X

These devices, connected in series with a transmission line, partially neutralize (or add to) the transmission line reactance. Therefore, they change the effective value of X in

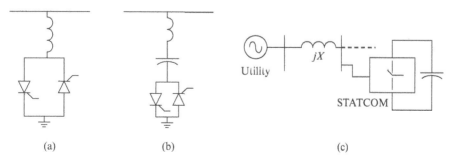

FIGURE 14.19 Shunt-connected devices for voltage control.

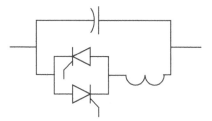

FIGURE 14.20   Thyristor-controlled series compensator (TCSC).

Equation (14.1), thus allowing the active power flow P to be controlled. Various forms of such devices have been used. These include the thyristor-controlled series capacitor (TCSC), also known as thyristor-controlled series compensator, shown in Figure 14.20. It should be noted that just as STATCOM with higher performance capabilities is a solid-state equivalent of an SVC, similarly, there is a solid-state equivalent of TCSC called SSSC (static synchronous series compensator), which injects a controlled voltage in series with the transmission line that either lags or leads the line current.

14.7.2.3   Static Phase Angle Control and Unified Power Flow Controller (UPFC)

Based on Equation (14.1), a device connected to a bus in a substation, as shown in Figure 14.21a, can influence power flow in three ways:

1. Controlling the voltage magnitude E,
2. Changing the line reactance X, and/or
3. Changing the power angle δ.

Such a device, called the unified power flow controller (UPFC), can affect power flow in any combination of the ways listed above. The block diagram of a UPFC is shown in Figure 14.21a at one side of the transmission line. It consists of two voltage-source switch-mode converters. The first converter injects a voltage $\bar{E}_3$ in series with the phase voltage such that

$$\bar{E}_1 + \bar{E}_3 = \bar{E}_2. \tag{14.2}$$

Therefore, by controlling the magnitude and the phase of the injected voltage $\bar{E}_3$ within the circle shown in Figure 14.21b, the magnitude and the phase of the bus

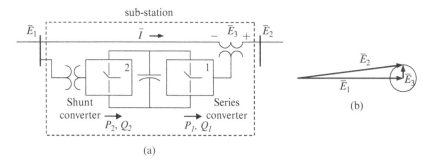

(a)

(b)

FIGURE 14.21   Unified power flow controller (UPFC).

voltage $\bar{E}_2$ can be controlled. If a component of the injected voltage $\bar{E}_3$ is made to be 90 degrees out of phase, for example leading with respect to the current phasor $\bar{I}$, then the transmission line reactance X is partially compensated.

The second converter in a UPFC is needed for the following reason: since converter 1 injects a series voltage $\bar{E}_3$, it delivers real power P_1, and the reactive power Q_1 to the transmission line (where P_1 and Q_1 can be either positive or negative):

$$P_1 = 3\,\mathrm{Re}(\bar{E}_3\bar{I}^*) \qquad (14.3)$$

$$Q_1 = 3\,\mathrm{Im}(\bar{E}_3\bar{I}^*). \qquad (14.4)$$

Since there is no steady-state energy storage capability within a UPFC, the power P_2 into converter 2 must equal P_1 if the losses are ignored:

$$P_2 = P_1. \qquad (14.5)$$

However, the reactive power Q_2 bears no relation to Q_1 and can be independently controlled within the voltage and current ratings of converter 2:

$$Q_2 \neq Q_1. \qquad (14.6)$$

By controlling Q_2 to control the magnitude of the bus voltage $\bar{E}_1$, UPFC provides the same functionality as that of an advanced static var compensator STATCOM. A UPFC combines several other functions: static var compensator, phase-shifting transformer, and controlled series compensation.

REFERENCES

1. N. Mohan, A. Jain, P. Jose, and R. Ayyanar, "Teaching Utility Applications of Power Electronics in a First Course on Power Systems," *IEEE Transactions on Power Systems* 19 (February 2004): 40–47.
2. N. Mohan, *First Course on Power Systems* (New York: John Wiley & Sons, 2011).
3. N.G. Hingorani, and L. Gyugyi, *Understanding FACTS* (New York,: IEEE Press, 2000).
4. D. Collins, "DOE FE Distributed Generation (DG) Fuel Cells Program," IEEE Applied Power Electronics Conference, March 6–10, 2005, Austin, TX.

PROBLEMS

14.1 Show the details and the average representation of converters in Figures 14.7b and 14.7c for wind-electric systems.

14.2 Show the details and the average representation of converters in Figure 14.8 for a photovoltaic system.

14.3 Show the details and the average representation of converters in Figure 14.11 for a flywheel storage system.

14.4 Show the details and the average representation of converters in Figure 14.12 for an adjustable-speed drive.

14.5 Show the details and the average representation of converters in Figure 14.14 for a UPS.

14.6 Show the details and the average representation of converters in Figure 14.15 for a DVR.

14.7 Show the details and the average representation of converters in Figure 14.16 for a current-link HVDC transmission system.

14.8 Show the details and the average representation of converters in Figure 14.17 for a voltage-link HVDC transmission system.

14.9 Show the details and the average representation of the converter in Figure 14.19c for a STATCOM.

14.10 Show the details and the average representation of converters in Figure 14.21a for a UPFC system.

INDEX

Power Electronics A First Course: Simulations and Laboratory Implementations, Second Edition.
Ned Mohan and Siddharth Raju.
© 2023 John Wiley & Sons, Inc. Published 2023 by John Wiley & Sons, Inc.
Companion Website: www.wiley.com/go/mohan/powerelectronics2e